FEBRUARY 2025

Critical Minerals and the Future of the U.S. Economy

EDITORS

Gracelin Baskaran

Duncan Wood

CONTRIBUTORS

Morgan Bazilian

Rohitesh Dhawan

Frank Fannon

Alexandra Helfgott

Adam Johnson

Joseph Majkut

Chris Michienzi

Jane Nakano

Meredith Schwartz

Seaver Wang

Kellee Wicker

Gregory Wischer

Matthew D. Zolnowski

An edited volume by the Critical Minerals Security Program

CSIS | CENTER FOR STRATEGIC & INTERNATIONAL STUDIES

BLOOMSBURY ACADEMIC

ABOUT CSIS

The Center for Strategic and International Studies (CSIS) is a bipartisan, nonprofit policy research organization dedicated to advancing practical ideas to address the world's greatest challenges.

Thomas J. Pritzker was named chairman of the CSIS Board of Trustees in 2015, succeeding former U.S. senator Sam Nunn (D-GA). Founded in 1962, CSIS is led by John J. Hamre, who has served as president and chief executive officer since 2000.

CSIS's purpose is to define the future of national security. We are guided by a distinct set of values—nonpartisanship, independent thought, innovative thinking, cross-disciplinary scholarship, integrity and professionalism, and talent development. CSIS's values work in concert toward the goal of making real-world impact.

CSIS scholars bring their policy expertise, judgment, and robust networks to their research, analysis, and recommendations. We organize conferences, publish, lecture, and make media appearances that aim to increase the knowledge, awareness, and salience of policy issues with relevant stakeholders and the interested public.

CSIS has impact when our research helps to inform the decisionmaking of key policymakers and the thinking of key influencers. We work toward a vision of a safer and more prosperous world.

Center for Strategic & International Studies
1616 Rhode Island Avenue, NW
Washington, DC 20036
202-887-0200 | www.csis.org

BLOOMSBURY ACADEMIC

Bloomsbury Publishing Inc, 1385 Broadway, New York, NY 10018, USA
Bloomsbury Publishing Plc, 50 Bedford Square, London, WC1B 3DP, UK
Bloomsbury Publishing Ireland, 29 Earlsfort Terrace, Dublin 2, D02 AY28, Ireland
BLOOMSBURY, BLOOMSBURY ACADEMIC and the Diana logo are trademarks of Bloomsbury Publishing Plc.

First published in the United States of America, 2025

Cover design: Leena Marte

Cover image © Adimas/Adobe Stock

ISBN: PB: 979-8-7651-9836-0
ePub: 979-8-7651-9837-7
ePDF: 979-8-7651-9838-4

Printed and bound in the United States of America

For product safety related questions contact productsafety@bloomsbury.com.

To find out more about our authors and books visit www.bloomsbury.com and sign up for our newsletters.

ACKNOWLEDGMENTS

The editors hope that this volume's findings, analysis, and policy recommendations will help inform the efforts of the U.S. presidential administration and Congress to build the secure and resilient minerals supply chains essential for U.S. national, economic, and energy security.

The editors would like to thank the authors and a series of individuals for their valuable contributions to this project. They would like to thank Meredith Schwartz for her exceptional editorial support and Caroline Smutny for her administrative support.

The editors would also like to thank Alex Kisling in the CSIS External Relations department; Phillip Meylan, Hunter Hallman, Kelsey Hartman, and Madison Bruno on the CSIS Publications team for their thoughtful feedback during the publication process; and Leena Marte and Will Taylor in the CSIS Dracopoulos iDeas Lab for bringing this to life for digital and print publication.

This volume is made possible through general support to CSIS and the Critical Minerals Security Program. No direct sponsorship contributed to this book. We are grateful to our donors, whose generosity made this important and timely undertaking possible.

CONTENTS

CHAPTER 1

Introduction

By Gracelin Baskaran

Yuichiro Chino via Getty Images

Mining is an inextricable part of the American story. What starts as rock in the ground goes on to become the inputs that build America's homes and buildings, transportation systems, energy generation and transmission, defense systems, and technological capabilities. Mining is the foundation that allowed the United States to be a military leader, providing the minerals needed to manufacture tanks, missiles, fighter jets and warships. It is the reason computers, phones, and iPads exist. Mining is the reason we have energy and can turn on lights every morning.

Today, the United States is 100 percent import reliant for 12 of the 50 minerals identified as critical by the U.S. Geological Survey (USGS) and over 50 percent import reliant for another 29. China is the top producer for 29 of these critical minerals.[1] This dominance is the result of decades of minerals-centered domestic industrial strategy and foreign policy. China has repeatedly shown its willingness to weaponize these minerals. Over the last two years, China has rolled out export restrictions, including complete bans, on antimony, gallium, germanium.[2] Furthermore, China has a stranglehold on minerals processing, refining between 40 and 90 percent of the world's supply of rare earth elements, graphite, lithium, cobalt, and copper.

Reducing reliance on China and creating resilient mineral supply chains is one of the most bipartisan priorities in Washington, D.C. This is demonstrated by the efforts of the last two administrations. In 2017, President Donald Trump issued Executive Order 13817, intending to improve the management of critical minerals needed for economic prosperity and energy security. In 2021, President Joe Biden issued Executive Order 14017, which led to a review of U.S. critical minerals and material supply chain vulnerabilities. The assessment released by the Biden administration discovered that the overreliance on adversarial countries posed a threat to national and economic security.[3]

Geopolitical tension and war have motivated the advancement of critical minerals policies for nearly a century. At the start of World War II, the United States adopted the Strategic and Critical Materials Stockpiling Act of 1939.[4] In his letter to Congress, President Franklin D. Roosevelt noted both that commercial stocks of vital raw resources in the United States were low and that "In the event of unlimited warfare on sea and in the air, possession of a reserve of these essential supplies might prove of vital importance."[5] By 1942, non-essential gold mining was restricted by the U.S. government so that it could free up mining companies' capacity to focus on

Figure 1: Share of Top Three Producing Countries in Mining of Selected Minerals, 2022

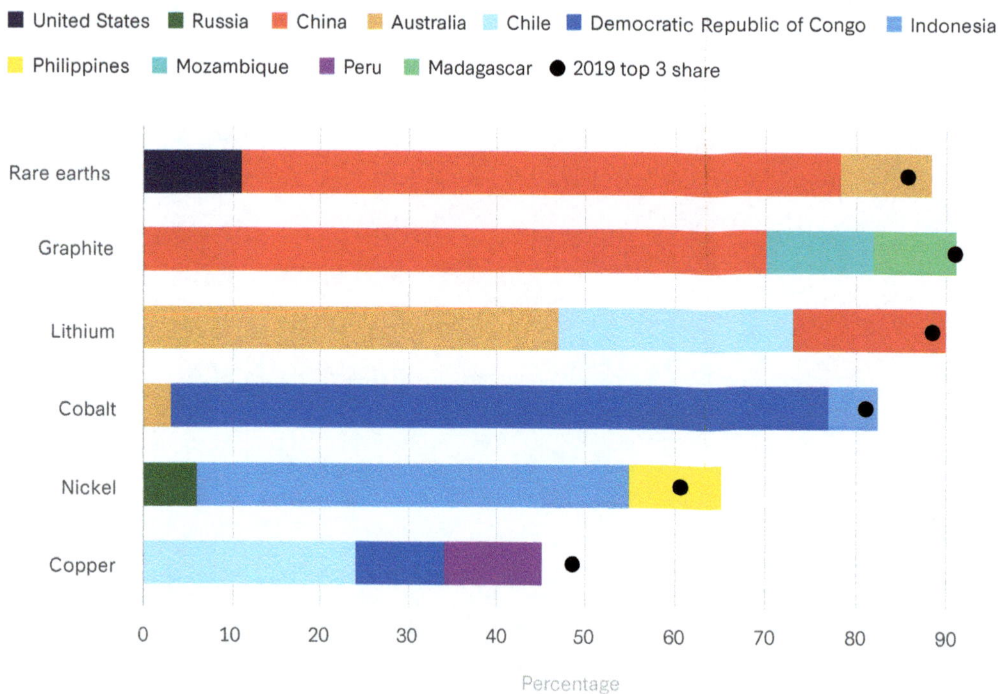

Source: This is a work derived by CSIS from IEA material and CSIS is solely liable and responsible for this derived work. The derived work is not endorsed by the IEA in any manner.

Figure 2: Share of Top Three Processing Countries of Selected Minerals, 2022

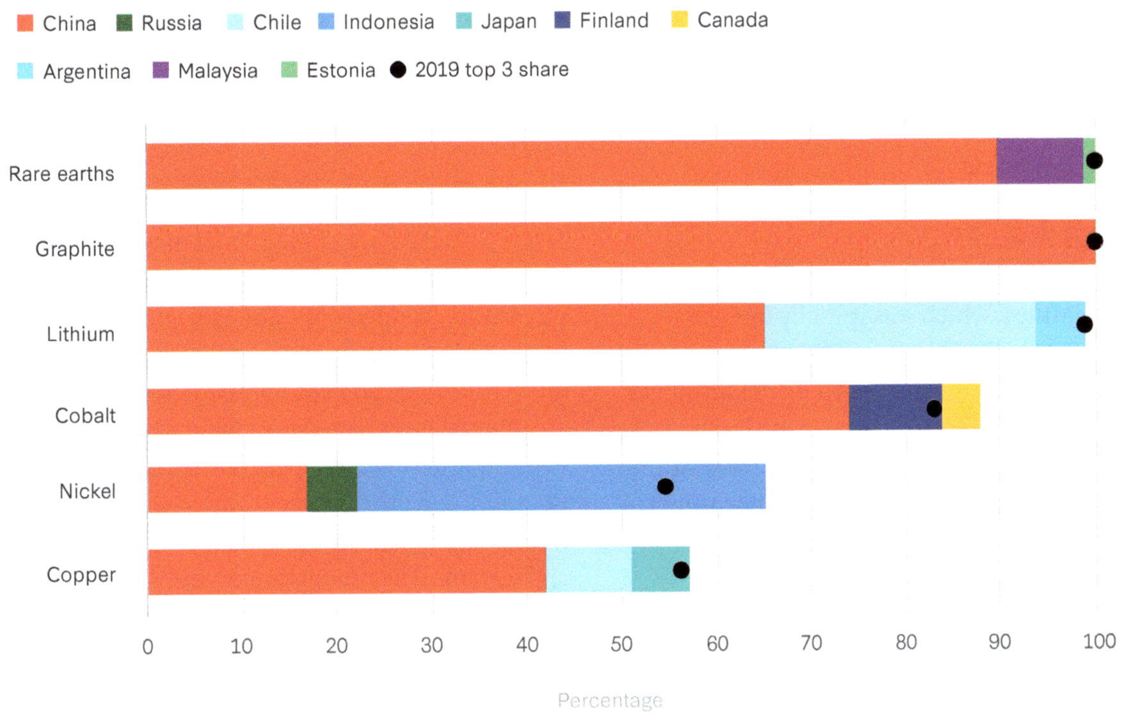

Source: This is a work derived by CSIS from IEA material and CSIS is solely liable and responsible for this derived work. The derived work is not endorsed by the IEA in any manner.

more critical minerals needed for the war effort.[6] Less than a decade later, the Defense Production Act of 1950 was passed in response to the Korean War and provided authority for allocations to source strategic minerals needed to manufacture defense technologies.[7]

While conflict and uncertainty have been the biggest drivers of advancing policies to build secure minerals supply chains, demand for minerals has largely been driven by industrialization, technological advancements, decarbonization, and economic growth. For example, in 1975, the United States required that catalytical convertors be installed into automobiles to reduce emissions. These catalytic convertors drove the long-term demand of platinum group metals and have single-handedly made American air cleaner and more breathable, reducing harmful exhaust gases from automobiles by over 90 percent.[8] Copper is another example. It is a necessary material for many of the advanced technologies that are essential to the modern global economy, including in infrastructure, clean energy, electronics, and automotives, and copper wires connect electrical grids, integrated circuits, and telecommunications systems. In order to meet net-zero carbon emissions by 2050, annual copper supply would need to double by 2035.[9] The artificial intelligence (AI) industry is putting additional pressure on copper demand. The data centers that process AI applications could demand up to 200,000 metric tons of copper per year between 2025 and 2028, adding another 2.6 million metric tons to the copper deficit in 2030.[10] Copper reached its highest ever price—$11,000 per metric ton—on the London Stock Exchange in 2024.[11]

As technologies advance and become cleaner, and as demand for them grows, the mineral needs of the U.S. economy intensify. The competitiveness of the U.S. domestic automotive, energy, technology, and defense industries will be key to determining the United States' standing as an economic powerhouse and global superpower in the decades ahead.

The United States will need to strengthen both its mission clarity and its execution. At present, the U.S. government has yet to agree on a single critical minerals list. Because copper is not on the USGS list, it has been ineligible for investment incentives from the Inflation Reduction Act (IRA). Additionally, there is no centralized agency or department to coordinate mining activities or execute a strategy. There are 15 government departments and agencies working on critical minerals, including the Departments of the Interior, Commerce, Energy, Defense, State, Labor, Homeland Security, Treasury, Agriculture, and Education; the Export-Import Bank; the U.S. International Development Finance Corporation; the U.S. Agency for International Development; the Environmental Protection Agency; and the National Aeronautics and Space Administration (NASA).[12] The Bureau of Mines, which was initially opened in 1910 to coordinate all mining activities, was closed in 1996 and never reopened. Most of these departments and agencies are working on their own critical minerals efforts, with little interagency collaboration. Ultimately, strengthening coordination within the U.S. government must be a priority.

This book has three sections. Section 1 provides an evaluation of the critical minerals needs of four vital industries—semiconductors, defense, electric vehicles, and renewable energy—and provides recommendations for strengthening the resilience of these supply chains. Section 2 evaluates key Biden-era initiatives related to minerals supply chains—the IRA, CHIPS and Science Act, Defense Production Act, and the Minerals Security Partnership—and provides recommendations for reforming. The final section provides an analysis of key issues—domestic permitting, building midstream processing capacity, commercial diplomacy for minerals, deep sea mining, responsible mining, and government coordination—and provides concrete recommendations for how the United States can strengthen its performance in these areas to be a competitive and credible global leader.

SECTION 1
CRUCIAL INDUSTRIES RELY ON CRITICAL MINERALS TO REMAIN COMPETITIVE

Safeguarding the supply chains for advanced technologies in strategic industries is an economic and national security imperative. Policymakers now face the immense task of fortifying supplies of everything from lithium and graphite for advanced battery chemistries to tungsten and rare earth elements for the next generation of warfighting technologies. Accordingly, the first part of this book delves into the mineral needs of four key industries to U.S. economic competitiveness: semiconductors, defense, automobiles, and renewable energy.

Semiconductors

Semiconductors are foundational to virtually every part of modern life, powering technologies that drive innovation, connectivity, and efficiency. They are used in smartphones, computers, military applications, medical devices, and automotives. Semiconductors are mineral intensive—small but essential quantities of gallium, germanium, palladium, silicon, arsenic, titanium, and other elements are needed to produce the array of semiconductors required for such diverse applications. The production of these resources is largely concentrated in foreign adversaries, exposing a severe national security risk. USGS has estimated that just a 30 percent supply disruption of gallium could cause a $602 billion decline in U.S. economic output, amounting to a 2.1 percent loss of gross domestic product (GDP)—a significant economic impact.[13] Semiconductor supply chains will not be secure until the necessary mineral supply chains are secured. In Chapter 2, Gracelin Baskaran and Meredith Schwartz assess the minerals needs of the semiconductor industry and provide recommendations on developing appropriate incentives and leveraging research and development.

Defense Industry

Minerals are the bedrock of the defense industry. They are used in a wide array of defense applications, including military weapons systems, ammunition, and aerospace technologies. China is rapidly investing in munitions and advanced weapons systems, acquiring new systems roughly five to six times more quickly than the United States.[14] While China is operating with a wartime mindset to enhance military readiness, the United States has maintained a peacetime approach. Even before new restrictions, the U.S. defense industrial base struggled with insufficient capacity and surge capability to meet production demands for defense technologies, many of which are highly minerals intensive. Restrictions on critical mineral supplies will further widen the gap, enabling China to advance its capabilities more effectively than the United States.

Figure 3: U.S. Import Reliance on China by Mineral Type (as Percent of Consumption)

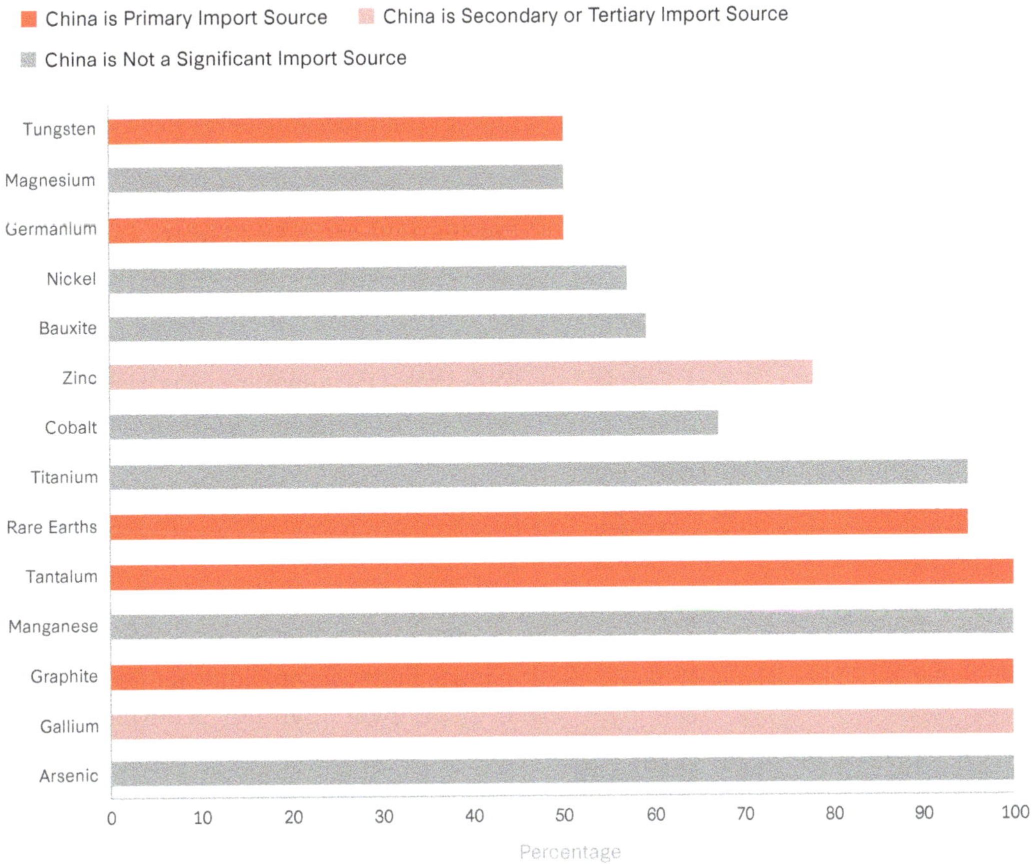

Source: USGS

In Chapter 3, Matt Zolnowski describes Department of Defense (DOD) actions to mitigate critical minerals vulnerabilities and advises how the department can update war-planning assumptions and stockpiling programs to prepare for a future crisis.

Electric Vehicles

Electric vehicles (EVs) are a major driver of innovation in the auto industry and are shaping the future of mobility. While the EV industry is important for the clean energy transition, it is also of vital importance to the U.S. economy. Domestic automakers began the commercial production of hybrid EVs in 1997.[15] They have spent decades investing in the development of the EV industry. EV investments in the United States over the last nine years reached $199 billion and created 201,900 EV-related jobs. By June 2024, automotive manufacturing jobs reached their highest levels since 1990.[16] EVs require significant quantities of minerals. While a traditional internal combustion engine (ICE) requires an average of 32 kg of critical minerals, an EV needs an average of 210 kg—over a six-fold increase. Therefore, the domestic auto industry now faces the daunting task of sourcing minerals from reliable and responsible partners amid a shifting policy landscape and swiftly evolving battery technologies. In Chapter 4, Duncan Wood and Alexandra Helfgott look at EV trends in the United States, assess the battery landscape, and examine how the United States should provide support to sustain an innovative and competitive domestic EV industry.

Renewable Energy

Renewable energy technologies will be key to unlocking new, cleaner sources of energy. In the United States, wind and solar together provide 15 percent of electricity generation, with both sectors poised for substantial growth. In 2023, there was $248 billion in clean energy investments. This is over a three-fold increase from 2018.[17] Today, the renewable energy industry employs 8 million people in the United States.[18] Southern states have been the biggest beneficiaries of clean energy investments, receiving $428 billion between the first quarter of 2018 and second quarter of 2024, followed by Western states ($327 billion), Midwestern states ($149 billion), and Northeastern states ($90 billion).[19] While this renewable energy buildout promises greater energy security, lower costs, and reduced emissions, it will depend heavily on secure mineral supplies. Wind farms and utility-scale solar facilities require far more mineral inputs than conventional power plants. Both technologies rely on critical minerals for their advanced electronics and components. In Chapter 5, Joseph Majkut highlights two materials especially imperiled due to rising demand and a lack of supply chain diversification—rare earth elements and polysilicon—which are essential for wind and solar power, respectively. China currently dominates the supply chains for both.

SECTION 2

PROGRESS UNDER THE BIDEN ADMINISTRATION

Important but Incomplete

In recent years, Washington has come to the realization that crucial U.S. industries and technologies are reliant on supply chains that are heavily concentrated in foreign adversaries, namely China. Advanced semiconductors designed in the United States are being manufactured and packaged in Taiwan. American automakers are producing EVs using Chinese battery chemistries. Domestically produced solar panels are made with Chinese solar cells and polysilicon. Over the course of decades and with the assistance of state-led industrial policies and billions of dollars in subsidies, China has grown to dominate the manufacturing sector for the cutting-edge technologies that the modern economy relies on. The risk this poses to U.S. national and economic security is untenable.

To that end, diversifying supply chains and boosting American manufacturing and ingenuity was a central objective of the Biden administration. To achieve these goals, President Biden enacted major pieces of legislation like the IRA and the CHIPS and Science Act, invoked the DOD's Defense Production Act to boost the industrial base, and initiated the Department of State's Mineral Security Partnership to further international cooperation.

These policy measures have been impactful in stimulating private sector investment as well as changing the narrative in policy circles around what qualifies as a strategic industry and what the role of government should be in securing supply chains. But these policy measures also have some significant gaps when it comes to the upstream mining and minerals industries. The new administration will be tasked with determining how these initiatives under the Biden administration can be modified, improved, and strengthened to better fit the mining industry's needs.

Inflation Reduction Act

The IRA is the Biden administration's flagship climate initiative, providing unprecedented levels of government incentives in the form of tax credits, grants, and loan guarantees for the clean energy industry. The bill includes provisions designed to address the entire circular clean energy supply chain, from the production of lithium and graphite to the manufacturing of EV batteries and wind turbines to the recycling and recovery of materials from end-use technologies. While the IRA has driven unprecedented investment in clean energy supply chains, sourcing critical minerals remains a critical limitation. In Chapter 6, Gracelin Baskaran and Meredith Schwartz score the IRA on how well it has achieved its objectives thus far and give recommendations as to how the minerals-related provisions may be altered and expanded upon to better meet the needs of industry and national security.

CHIPS and Science Act

The CHIPS Act, signed into law in August 2022, was an amalgamation of a number of legislative efforts to address both a rising China and the United States' desire to firm up access to semiconductors following the pandemic supply chain shock that froze consumer access to a wide range of products. Although this bill addressed a wide range of science and technology areas relevant to U.S. competitiveness, it did not prioritize securing American access to critical minerals. As a result, the CHIPS Act's focus on critical minerals access is minor, with the majority of funding going toward research and development for chip innovation, workforce development programs, and, above all, attracting investment in semiconductor fabrication, assembly, testing, and advanced packaging. Congress needs to act to explicitly address minerals relevant to semiconductor production, as they have for minerals used for EVs and clean tech. In Chapter 7, Kellee Wicker analyzes the impacts and shortcomings of the CHIPS Act and provides recommendations for strengthening the legislation.

Defense Production Act

The Biden administration also invoked the Defense Production Act (DPA) of 1950 to secure critical minerals for the defense industrial base. The DPA authorizes the president to ensure the availability of U.S. and Canadian industry for U.S. defense, essential civilian, and homeland security requirements. DPA Title III, Expansion of Productive Capacity and Supply, includes incentives for the DOD to develop, maintain, modernize, and expand production capacity or critical technologies. DPA Title III funds cannot be used if other funding (e.g., private investment or funding from other agencies) can be secured. Given the private sector's reluctance to make investments in critical minerals projects due to market price volatility for these materials, the DPA has proven to be a vital financing mechanism. In Chapter 8, Christine Michienzi details how the DPA has been used so far to support the critical minerals industry in the United States and Canada and gives recommendations as to how the new administration can best leverage the program.

Minerals Security Partnership

The Minerals Security Partnership (MSP) is a multilateral State Department initiative uniquely focused on minerals security. Since its inception in 2022, the MSP has mobilized a coalition of market-led democracies, primarily Western developed nations, with India as the only developing nation with membership. By 2024, the MSP supported nearly 30 minerals projects around the world and has brought additional mineral-rich countries to the table in the MSP Forum. But many unknowns and questions still remain as to how efficacious the MSP is and how it endures in a new administration. In Chapter 9, Jane Nakano suggests several modifications to the MSP that could accord more dynamism and long-term durability.

SECTION 3

ADDRESSING CHALLENGES AND OUTSTANDING QUESTIONS IN THE CRITICAL MINERALS INDUSTRY

The minerals industry faces a number of unique challenges that policymakers must address in the coming years if they wish to substantially shift mineral supply chains and improve U.S. industry's access to non-Chinese mineral sources. The third part of this volume delves into some of the biggest issues and questions facing the industry, from how to expedite the domestic mine permitting process to whether deep sea mining is the future of minerals extraction.

Domestic Permitting

Mining is the first step in the critical minerals supply chain, yet permitting a mine is a major hurdle in domestic critical mineral production that has yet to be overcome. On average, it takes 29 years to build a mine in the United States, the second-longest time in the world. Obtaining permission to operate a mine in the United States today involves securing federal, state, and local permits. A project can require up to 30 permits, many of which are duplicative.[20] Policymakers on both sides of the aisle are calling for a modernized permitting system that facilitates the development of domestic mining projects. In Chapter 10, Morgan Bazilian and Gregory Wischer review the history of permitting policy in the United States and provide actionable policy solutions to streamline the process.

Building Processing and Refining Capacity

In the next stage of mineral production, also known as the midstream, mined mineral ore must be processed and refined into the high-purity metals and materials used in end products. This stage of the supply chain is where China truly dominates. The dependence on Chinese processing creates strategic vulnerabilities, exposing the United States to potential supply disruptions due to geopolitical tensions, export restrictions, and price manipulations. To reduce these risks and bolster national security, it is essential to enhance U.S. midstream processing capabilities. In Chapter 11, Adam Johnson explains the importance of building domestic mineral processing capacity and provides recommendations on developing the workforce, leveraging strategic reserves, and streamlining permitting to accelerate the development of midstream capabilities.

International Engagement

Over the past 30 years, China has emerged as a key player in mineral supply chains through strategic international engagement. Although it produces only 10 percent of the world's lithium, cobalt, nickel, and copper, China imports sufficient quantities to process 65 to 90 percent of the global supply of these metals. This dominance is the result of years of industrial planning and foreign policy initiatives from Beijing. Given the United States' limited domestic reserves—including less than 1 percent of the world's reserves of commodities such as cobalt, nickel, and graphite, and less than 2 percent of manganese and rare earth elements—it must develop a strategy to reduce its dependence and enhance its mineral supply security. In Chapter 12, Gracelin Baskaran provides a novel framework for determining which international partners to prioritize and gives recommendations for how policymakers should engage in commercial diplomacy. These efforts should prioritize financing minerals security needs, leveraging soft power through infrastructure development and geological mapping, and developing carrots and sticks to drive market-based activity aligned to U.S. government interests.

Deep Sea Mining

While today's EVs and semiconductors are manufactured with minerals from land-based mines, this may not always be the case. Minerals that are found in nodules at the depths of the ocean, including manganese, nickel, copper, and cobalt, offer immense untapped resource potential. However, the environmental impacts of extraction from these sources remain largely unknown, and a set of international regulations has yet to be finalized. Still, China continues to make strides in developing the necessary technologies and exploration licenses to capture deep sea resources first. In Chapter 13, Seaver Wang provides insight into the status of the deep sea mining industry and the opportunity to change the calculus around mineral supply chains by expanding tax incentives to cover minerals mined from the sea, strengthening support to seafloor minerals research to improve environmental management approaches, and providing financing to strategic demonstration projects.

Responsible Mining

Mining is an industry with a complicated, and often negative, reputation due to all-too-frequent incidents of environmental degradation, human rights violations, social unrest, and devastating workplace accidents. Therefore, responsible mining standards are a nonnegotiable to ensure that U.S. and allied mines operate under best practices. As a major consumer and increasingly important producer of mined materials, the United States has a critical role in promoting responsible mining practices. In Chapter 14, Rohitesh Dhawan offers insight into how permitting reform can be done in a way that promotes responsible mining practices and advantages projects that follow high standards, how responsible mining standards can be used as a criterion for public procurement of metals or metal-based products, and how green premiums can be leveraged to financially incentivize responsible mining.

A Comprehensive U.S. Strategy for Minerals Security

Minerals policy has shifted quite significantly since 2010. The Obama, Trump, and Biden administrations all approached the critical minerals challenge based on their respective times and the policy tools at their disposal. America has learned much from these experiences. Reflecting on those experiences, Frank Fannon provides a suite of recommendations in Chapter 15 for the new administration to retake the commanding heights of the new economy: developing a single point of accountability to oversee and coordinate the administration's multiple lines of minerals policy efforts, reforming financing tools such as the DFC and Export-Import Bank (EXIM), undertaking permitting reform, and eliminating provisions that allow firms with any Chinese ownership from receiving taxpayer subsidies. Successful action will require an "all-of-the-above" approach.

Looking Ahead

The mission we undertook in compiling this book was to provide a comprehensive analysis of the indispensable role critical minerals play in the modern economy's most strategic industries, and to more fully understand and address the vulnerabilities the United States faces in securing the minerals upon which it so clearly depends. Furthermore, this volume is rich in policy proposals for the new administration, laying out a path forward for the most pressing challenges facing the critical minerals supply chain, from extraction to processing and refining to end use. These challenges are real and profound and require urgent attention—but as the chapters in this book demonstrate, they are not insurmountable.

SECTION 1

Crucial Industries Rely on Critical Minerals to Remain Competitive

CHAPTER 2

Powering Technology

Critical Minerals for the Semiconductor Industry

By Gracelin Baskaran and Meredith Schwartz

Narumon Bowonkitwanchai via Getty Images

The success of the Western semiconductor industry depends on reliable access to the critical minerals that are responsible for continuous advancements in the industry.

Semiconductors are the fundamental building blocks of modern technology, necessary for everything from smartphones and laptops to communications and energy-storage systems to military and aerospace applications.[21] These integrated circuits are called "semiconductors" due to being partial conductors, a unique property that enables them to control the flow of electrons by acting as both conductors and insulators.[22] Therefore, semiconductors rely on small but essential quantities of specific minerals with these properties to function.

Silicon, gallium, and germanium are the most common semiconductor materials used to form wafers, with different chip applications calling for different materials.[23] However, a myriad of other critical minerals come into play during the manufacturing stages and doping process—in which additional metals are introduced to slightly alter the chip's conductivity—to create just one integrated circuit.[24] Palladium, arsenic, iridium, titanium, copper, and cobalt are just some of the additional minerals that are necessary for semiconductor plating, wiring, doping, and packaging during production.[25]

The critical minerals most central to semiconductor production have high-risk supply chains largely concentrated in China. China produces 98 percent of the world's refined gallium and controls 68 percent of refined germanium production, 79 percent of the world's silicon, 40 percent of its arsenic trioxide, and 67 percent of its titanium.[26] The United States, meanwhile, is reliant on imports to access the materials needed for high-performance semiconductors. In 2022, the United States produced no arsenic, no gallium, less than 2 percent of the world's refined germanium, 3 percent of its silicon, and less than 1 percent of its titanium.[27]

The concentration of global critical minerals supply chains in the hands of adversaries presents a major security challenge for Western chipmakers. The chance of prolonged and widespread supply disruptions for semiconductor minerals is high, as China has already demonstrated its ability to restrict the flow of key minerals in the global economy. In July 2023, China announced export restrictions on gallium and germanium.[28] A year and a half later, China cut off the United States from Chinese gallium and germanium entirely through complete export bans on these materials targeted specifically at the United States.[29]

The semiconductor industry is too central to U.S. economic and national security to allow such an evident vulnerability in its supply chain. To truly secure the Western semiconductor industry, policymakers should address mineral supply chain vulnerabilities, not just vulnerabilities in downstream chip manufacturing.

THE IMPORTANCE OF MINERALS TO ADVANCE SEMICONDUCTOR TECHNOLOGY

The critical minerals necessary for semiconductor production hold the key to furthering innovation in the industry. A concept commonly known as Moore's Law stipulates that the density of a semiconductor (i.e., the number of transistors that can fit on a 1-square-inch microchip) will continue to rise every year, equating to more computing power, higher speed, and more complex applications.[30] For decades, silicon has been the wafer material of choice for most semiconductors due to its abundance in nature and thermal stability, making it a cheaper choice well suited to the early electronics industry. However, silicon alone may be close to reaching the physical limits of Moore's Law.[31] Rather, gallium and germanium are essential additions and alternatives to unlocking more advanced chipmaking.

Gallium and germanium have certain advantages over silicon that make them ideal materials for increasingly advanced semiconductors. Germanium's high electron mobility allows it to conduct electrons nearly three times faster than silicon, translating into faster device performance.[32] Semiconductors with germanium channels, known as complementary metal-oxide semiconductor (CMOS) circuits, are used today for quantum computers. Gallium similarly offers greater conductive potential for higher power density and energy efficiency.[33] High-performance chips made with gallium nitride (GaN) and gallium arsenide (GaAs) are used in advanced defense applications from satellite communications to missile detection systems. Production of GaN chips is expected to grow more than 25 percent annually through 2030, with defense applications driving this increase.[34]

Gallium and germanium are indispensable materials for the future of the semiconductor industry. But with current supply chain challenges and no U.S. sourcing alternatives, these materials are increasingly difficult to obtain. The next generation of chipmaking requires policymakers to devise and execute a critical minerals strategy that ensures the industry will have a reliable supply of needed materials.

MINERAL SOURCING CHALLENGES

Gallium and germanium are especially rare in the Earth's crust, at only 19.0 and 1.6 parts per million (ppm), respectively. Copper, in comparison, is estimated at 60 parts per million.[35] These concentrations of minerals are too widely dispersed to be recovered directly from the Earth. Rather, the only economically viable way to source gallium and germanium is to recover them as byproducts from the mining and processing of other minerals. Gallium is sourced from bauxite ores through aluminum smelting, and germanium is primarily recovered from zinc smelting. Even so, less than 10 percent of the gallium in bauxite and 5 percent of the germanium in zinc can be recovered.[36] These materials must then go through a complex refining process to produce gallium and germanium at the needed purity levels of over 99.99 percent.[37]

China has several advantages in gallium and germanium sourcing. The country has rich zinc deposits and imports 75 percent of the world's bauxite due to its leading aluminum industry.[38] This access to feedstock has also positioned China well to lead in germanium and gallium recovery. Aided by government subsidies, Chinese firms were able to flood the market with mineral oversupply in the 2010s.[39] As prices dropped, Western competitors could not remain economically viable, allowing China to emerge as the world leader in semiconductor minerals.[40]

In contrast, the United States has small bauxite reserves of 20 million tons (less than 1 percent of global totals) and limited zinc reserves of 76.6 million tons (3 percent of global totals).[41] The country currently has limited mining activity and produces only small amounts of germanium from zinc deposits in Alaska and smelting operations in Tennessee. Some new domestic projects may be in the works: In 2023, Dutch company Nyrstar announced a $150 million investment to expand its existing zinc operations in Tennessee to add a gallium and germanium processing facility.[42] However, the project has yet to secure investor funding, and the company has faced market challenges that led it to temporarily suspend zinc mining operations in October 2023.[43] In the near term, domestic investments will evidently not be enough to secure gallium and germanium supply chains.

U.S. allies and strategic partners will be key to sourcing bauxite and zinc and producing gallium and germanium. For example, although Australia is the top producer of bauxite and home to the largest zinc reserves in the world, it lacks midstream processing capacity, leading it to send over 50 percent of its zinc exports and 97 percent of its bauxite exports to China.[44] And Peru, a U.S. free trade partner with the largest zinc smelting plant in Latin America, currently produces no germanium or gallium.[45] Australia and Peru hold vast potential for alternative gallium and germanium sourcing for the semiconductor industry, but without investment by Western firms into midstream processing and refining, these resources will remain untapped.

THE CHIPS AND SCIENCE ACT

But What About the Critical Minerals?

In the spring of 2020, at the peak of the Covid-19 pandemic, the United States experienced firsthand how debilitating semiconductor shortages can be. An estimated 169 sectors and consumer lines were impacted by semiconductor supply disruptions, including the electronic, automotive, communications, and healthcare industries. As a result, Western firms faced lower production volumes, the cancellation of new product lines, and delayed breakthroughs in technologies such as artificial intelligence (AI) and the Internet of Things (IoT).[46] Policymakers realized just how fragile current semiconductor supply chains are, due to a highly complex and specialized production process largely concentrated in Asia.

In August 2022, President Joe Biden signed the Creating Helpful Incentives to Produce Semiconductors (CHIPS) and Science Act into law, with the goal of strengthening U.S. semiconductor manufacturing and supply chains. This act included over $280 billion in support for advanced chip manufacturing, packaging, and workforce development.[47] To address the supply chain vulnerabilities experienced during the pandemic, the bill focused on onshoring downstream capabilities, including by developing fabrication facilities for legacy chips used in communications and defense applications.[48] The legislation also introduced significant government grant funding, which has been awarded to companies such as Intel and Micron to enable them to build and expand their chip-manufacturing capacity.[49] However, the CHIPS Act overlooked a major national security vulnerability in semiconductor supply chains: critical minerals. The bill did not include any provisions to incentivize the diversification of critical minerals supply chains for semiconductors.

EXPORT RESTRICTIONS HIGHLIGHT SUPPLY CHAIN VULNERABILITIES

The United States quickly realized just how big an oversight the exclusion of critical minerals from the CHIPS Act was. On August 1, 2023, Chinese export restrictions on gallium and germanium went into effect in retaliation for Washington banning exports of advanced semiconductor technologies to China.[50] Due to the restrictions, gallium and germanium exporters in China are now required to apply for an export license for each shipment of material, providing the government with details on the overseas buyer and end use.[51] Beijing justified the escalation of the tech war by claiming that such measures were necessary for national security. Meanwhile, the Western semiconductor industry was paying the price.

In August and September 2023, China exported no refined gallium and only 1 kg of refined germanium, compared to nearly 8,000 kg and 6,900 kg, respectively, in the preceding July.[52] In total, China's gallium exports for 2023 were over 50 percent lower than exports for 2022; as of February 2024, gallium exports had yet to return to pre-restriction levels, and it remains unclear when China's exports will return to their previous peaks.[53] Continued restrictions and the implementation of gallium and germanium export bans in January 2025 will have a significant effect on the U.S. economy. The U.S. Geological Survey (USGS) has estimated that a disruption to just 30 percent of gallium supply could cause a $602 billion drop in U.S. economic output, equivalent to 2.1 percent of gross domestic product.[54]

Prices for these materials have risen markedly over the past year. In April 2024, gallium prices were at their highest level since 2011. Assessed prices for gallium have nearly doubled since the restrictions were imposed, and germanium prices have also climbed over 70 percent, to $2,280 per kilogram.[55] China has demonstrated its ability to control the materials market for semiconductors and has only tightened U.S. access to these materials with newly implemented exported bans. Continued and additional bottlenecks in critical minerals supply present an ongoing threat to the resilience of the U.S. semiconductor industry.

RECOMMENDATIONS

Creating Better Policy for Semiconductor Mineral Supply Chains

The CHIPS and Science Act, as well as the current policy focus on downstream chip manufacturing, will not be enough to secure semiconductor supply chains. As long as China controls critical minerals supply chains, the U.S. semiconductor industry will be vulnerable to export restrictions, bottlenecks, and price volatility.

One policy recommendation that is frequently cited as a solution to shortages of base metals is to revamp government stockpiling efforts through the National Defense Stockpile under the Strategic and Critical Materials Stock Piling Act of 1939. The United States currently stockpiles no gallium and only about 14,000 kg of germanium, or half of the country's annual consumption.[56] However, stockpiling semiconductor metals to address critical minerals supply chain vulnerabilities will be challenging due to their price volatility and the small quantities needed for the industry (relative to the electric vehicle industry, for example, which requires significant quantities of base metals).[57] In addition, gallium has a shelf life of only around one year due to its low melting point.[58] Stockpiling efforts are therefore not the best solution to addressing supply chain concerns.

Instead, the United States should consider the following actions:

- **Invest in building the technological know-how for gallium and germanium refining.** Refining these minerals to needed purity levels of over 99.99 percent for the semiconductor industry requires specific technology, infrastructure, and expertise, all of which are currently lacking. The United States has only one company that refines high-purity gallium and one operation for germanium.[59] A research and development laboratory could boost innovation to increase processing capacity and produce minerals in a more cost-effective way.

  The Department of Energy already funds laboratories focused on critical minerals for electric vehicles (EVs) and clean energy, but there is less focus on semiconductor minerals such as gallium and germanium. For example, the Critical Minerals Innovation Hub at the Ames National Laboratory in Iowa and the Minerals to Materials Supply Chain Research Facility (METALLIC) network bring together the expertise of several leading national laboratories to find solutions to critical minerals supply challenges for clean energy industries.[60] The Department of Commerce should fund similar initiatives focused on minerals for the semiconductor industry. National laboratories can help develop the capabilities, technologies, and skills needed to produce refined semiconductor minerals at scale.

- **Put together a comprehensive incentives package that addresses upstream critical minerals mining and midstream processing and refining.** Just as the CHIPS and Science Act focused on incentives to boost domestic semiconductor manufacturing and the Inflation Reduction Act (IRA) incentivized investments in EV and clean energy technologies, an upstream and midstream critical minerals incentives package would ensure that U.S. fabs manufacturing the next generation of semiconductor technologies have reliable access to the critical minerals they need. With billions of dollars being invested in chips foundry facilities, their success hinges on access to needed input materials.[61] Ensuring the security of the critical minerals supply chain is common-sense policy that supports the ambitious industrial goals of the CHIPS and Science Act and IRA.

  This package should include investment and production tax credits such as those covered in Sections 48C and 45X of the IRA. Such incentives programs would encourage companies to make the necessary investments in critical minerals recovery and refining facilities amid uncertain and volatile market conditions. Midstream projects like Nyrstar's gallium and germanium recovery plant are struggling to secure financing in the face of steep competition from Chinese firms that have a history of pricing out Western competitors.[62] Federal tax credit programs signal to the private sector that the government is supportive of the industry and offer an additional cash incentive boost to projects that may otherwise stall. These incentives should apply to both domestic projects as well as ones in strategic allied countries that have high potential for gallium and germanium production, such as Australia and Peru.

  An incentives package should also include grant funding similar to the large dollar amounts currently being awarded by the CHIPS Program Office within the Department of Commerce for onshoring semiconductor fabs. The U.S. government can incentivize mining companies to make significant investments in gallium and germanium recovery and refining facilities and infrastructure by alleviating some of the capital burden. Just as semiconductor manufacturing facilities require an immense amount of capital, standing up domestic gallium and germanium mining, processing, and refining infrastructure will require massive investment. These projects are not only essential to boosting supply chain security for the high-tech industries downstream that depend on critical minerals, but they will also create jobs, onshore manufacturing capacity and niche skills, and help revitalize mining communities that have been economically left behind.

CONCLUSION

The success of the Western semiconductor industry depends on reliable access to the critical minerals that are responsible for continuous advancements in the industry. The CHIPS and Science Act of 2022 sought to build a domestic ecosystem for a thriving semiconductor industry that is invulnerable to the supply chain risks of the past. But this strategy was incomplete, as there has been no U.S. policy to date addressing the mineral needs of chips manufacturers. As China imposes mineral export restrictions and squeezes supply, policymakers can no longer afford to overlook mineral security. A comprehensive incentives package is needed to build research and development institutions and boost the upstream and midstream capacity needed to onshore and friend-shore gallium and germanium production.

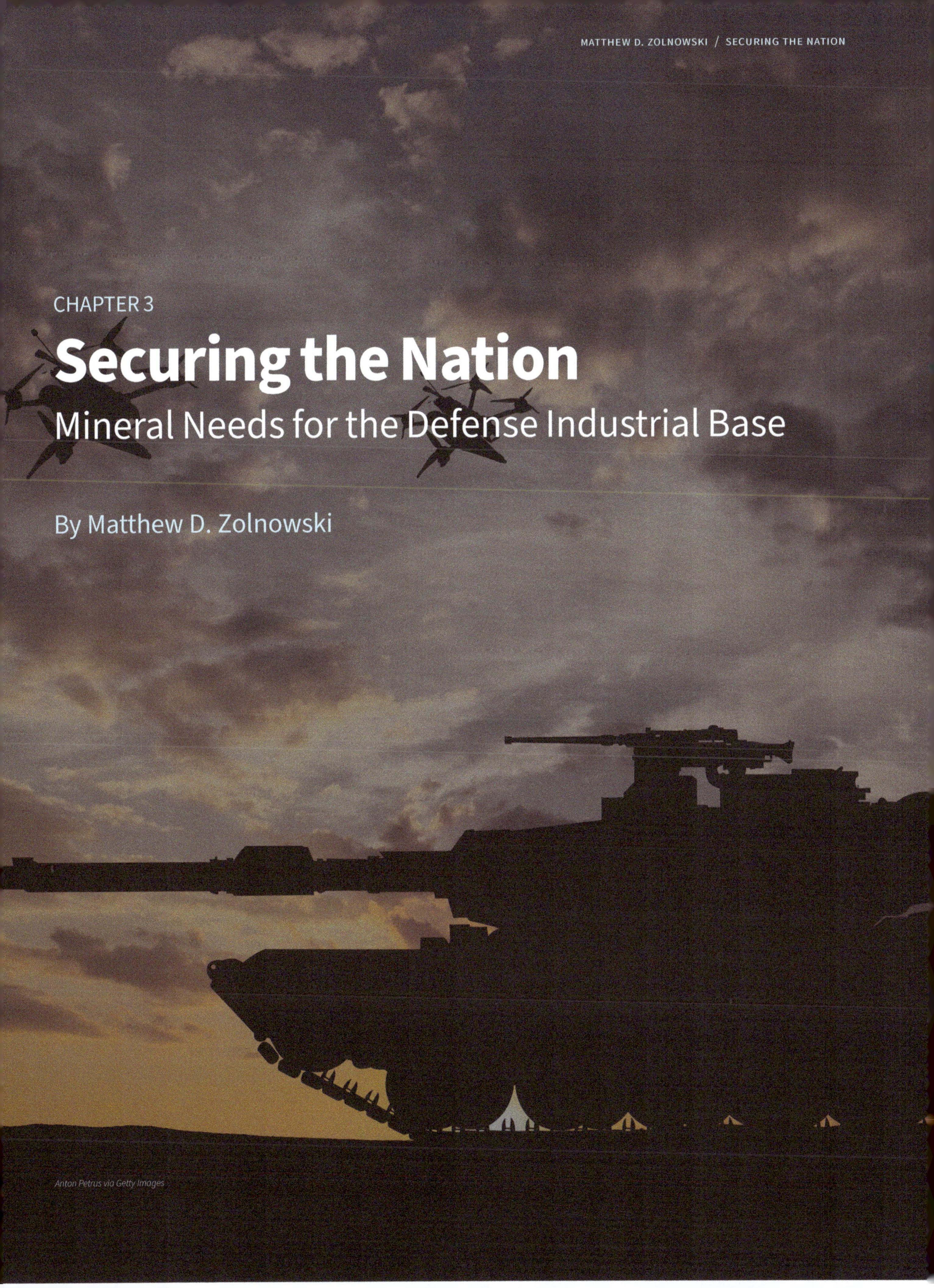

CHAPTER 3

Securing the Nation

Mineral Needs for the Defense Industrial Base

By Matthew D. Zolnowski

Anton Petrus via Getty Images

The defense industrial base is at risk of critical minerals shortages in an emergency, with industrial mobilization doctrine and program execution mired in a peace dividend posture.

As the furthest upstream tier of defense supply chains, critical minerals support virtually all Department of Defense (DOD) activities and platforms, whether through indirect consumption, such as a rare earth catalyst for petroleum refining, or direct consumption, such as aluminum and titanium parts in an aircraft. In some cases, the critical minerals supporting a defense system are indistinguishable from those used in civilian products; in others, critical minerals are converted into military-unique formulations that enable a weapon system's cutting-edge performance.

However, this critical minerals consumption pattern is a constant for all military organizations throughout human history—whether using bombs and bullets, shot and pike, or sling and stone. Therefore, incorporation into a weapon system is not, by itself, sufficient reason for critical minerals to be deemed essential to national defense. This type of usage certainly would not justify the DOD's deployment of over $1 billion since 2019 under the Defense Production Act (DPA) of 1950 and other authorities to expand domestic and allied production of critical minerals.

With this contradiction in mind, this chapter aims to describe the process by which the DOD determines whether a critical mineral rises to the level of a national defense requirement, as well as how the DOD and industry are addressing such needs.

In its simplest form, the DOD's assessment of the "criticality" of a critical mineral is directly connected to the National Defense Strategy and its policy assumptions about the conflicts in which the DOD may be called to fight. Though defense planners historically focused on protracted conflict, the DOD has drifted toward a more optimistic policy baseline: a single-year conflict followed by a multiyear reconstitution period.

Even under this more optimistic baseline, the DOD has identified significant supply deficits to defense requirements during a national emergency scenario, covering 69 materials and valued at $2.41 billion.[63] In addition to the large-scale industry investments previously noted, these findings have prompted the DOD to embark on wide-ranging reforms to its critical minerals stockpiling law, as well as to tighten procurement restrictions to reduce reliance on adversarial nations for critical minerals.

Though the DOD has made significant strides in modernizing statutory authorities and deploying an array of programs to address its critical minerals needs, many of its planning processes related to requirements generation and industrial mobilization remain rooted in the immediate post–Cold War period. Maintaining the positive progress to date, while reviewing and updating those policies and programs that have not kept pace, should be the DOD's next area of focus.

THE SPECTRUM OF MILITARY ACTIVITIES REQUIREMENTS GENERATION

The spectrum of activities undertaken by the U.S. Armed Forces is vast. They conduct military-to-military diplomacy, peacekeeping operations, and potential combat operations, ranging from raids by special operations forces to the deployment of a multinational force for large-scale conventional war.

Amid this extreme variability, the DOD has developed a structured process to collect critical minerals data and evaluate which minerals are necessary for both essential civilian and defense industries across a range of scenarios. The DOD has made no public report of its critical minerals needs since 2015, but the results of its assessments are disseminated across the U.S. government every two years.[64]

At a high level, this process begins with the collection of economic and technical data related to critical minerals markets. This data is integrated by the Defense Logistics Agency (DLA) Strategic Materials, the administrator of the National Defense Stockpile (NDS), for analysis through a series of economic models. These models project the anticipated supply and demand of critical minerals over a given period, after which supply and demand are perturbed.

These perturbations are driven by policy judgments related to the execution of a military conflict scenario used for DOD budgetary and planning purposes—underpinned by the National Defense Strategy. The elements described in this scenario include the following:

- the duration of the conflict
- the military force deployed
- combat losses
- military, industrial, and essential civilian demand
- shipping losses
- the availability of foreign supply
- domestic industrial mobilization
- civilian austerity measures[65]

Each of these elements is highly subjective, and historically these parameters have been hotly debated between defense planners concerned about a protracted war and those who expect conventional wars involving the U.S. Armed Forces to end quickly.[66]

In its Cold War iteration, advocates of a "short war" planning construct argued that a potential conflict between the United States and the Soviet Union would be extraordinarily violent in its initial phases or might rapidly escalate to the nuclear level. In its contemporary iteration, advocates note the overwhelming conventional advantage of the U.S. Armed Forces over most threats—ably demonstrated in the First Gulf War—suggesting conflicts involving U.S. conventional forces are likely to be very short.

In either case, industrial preparedness and stockpiling of any kind would be unnecessary, as the conflict may end before these efforts could impact its outcome.

In contrast, the legacy "long war" proponents argue that an asymmetry of conventional military power is an unreliable indicator of conflict duration—ably demonstrated by multi-decade counterinsurgency campaigns—and that an industrial base specialized toward a small, highly sophisticated fighting force will struggle to grow in a protracted conflict.

In the "long war" argument, important factors such as where and how U.S. Armed Forces may fight remain highly uncertain. Ultimately, this uncertainty drives hedging behavior, using stockpiles to mitigate demand for conflict surge items or minerals until new wartime production can come online.

From the initial promulgation of the Strategic and Critical Materials Stock Piling Act, the combination of World War II and Korean War experiences led U.S. government planners to favor a "long war" planning construct. A five-year war scenario drove the creation of large NDS inventories and broad industrial mobilization activities under the DPA. As the Cold War progressed, subsequent administrations embraced more optimistic war-planning and economic policy judgments, each driving the U.S. government toward smaller critical minerals stockpiles and industrial preparedness efforts.[67]

The "short war" planning construct adopted at the end of the Cold War remains the DOD's baseline for critical minerals requirements generation today—a one-year conflict, followed by three years to reconstitute the U.S. Armed Forces.[68] Based on these results, the DOD has implemented an array of critical minerals mitigation programs, principally focused on aerospace, operational energy, and armor needs (see Table 1).[69]

Notwithstanding the breadth of critical minerals mitigation programs underway, the unclassified summary of the "base case" results from the Strategic and Critical Materials 2023 Report on Stockpile Requirements identifies ongoing critical minerals requirements in a national emergency. This recent DOD study identified shortfalls to defense requirements for 69 materials, valued at $2.41 billion, and shortfalls to essential civilian demand for 24 materials, valued at approximately $12.21 billion.[70]

More simply, these results suggest that substantial portions of the DOD's "short war" critical minerals needs remain unaddressed.

Table 1: Critical Minerals Mitigation Programs

Mineral	Sample Use Case
Aluminum	Aerospace alloys (lightweighting), armor
Antimony	Ammunition, fire retardants
Beryllium	Aerospace alloys (lightweighting, non-sparking)
Boron	Armor
Cobalt	Batteries, aerospace alloys (engines)
Germanium	Space-based solar cells
Graphite	Batteries
Lithium	Batteries
Magnesium	Aerospace alloys (lightweight)
Manganese	Batteries
Nickel	Batteries
Niobium	Aerospace alloys (engines), shipbuilding steel
Steel	Armor
Tin	Electronics
Titanium	Aerospace alloys (structural and engines)
Tungsten	Ammunition, cutting implements
Rare earth elements	Control and actuation systems, ceramic materials

Source: Author's analysis of awards under Catalog of Federal Domestic Assistance (CFDA) 12.777 Defense Production Act Title III, retrieved from USASpending.gov.

GOVERNMENT AND INDUSTRY MITIGATION TOOLS

Given the massive gap between current industrial base capabilities and postulated DOD requirements, the DOD pulls multiple levers to address day-to-day and future planning for critical minerals supply chain risk mitigation. Many of these levers are described in Joint Publication 4-05: Joint Mobilization Planning (JP 4-05), and in its most recent iteration, the principal industrial mobilization tools listed by the DOD include the following lines of effort:

- actively employ DPA Title I to prioritize DOD contracts or allocate scarce materials to defense contracts
- expand military production and supporting sectors (e.g., workforce development)
- draw upon Canadian industrial capacity to supplement U.S. production
- obtain other allied weapons production support
- obtain waivers or exemptions from U.S. environmental laws to facilitate the above[71]

However, just as planning assumptions for NDS functions have remained on autopilot since the end of the Cold War, the industrial base mitigation tool kit—and the DOD doctrine governing it—has remained largely unchanged. More specifically, JP 4-05 is almost word-for-word identical from 1995 to the present, and shortcomings in doctrine and practice have yet to be addressed.

Contracting

DPA Title I requires a U.S. company, and any of its suppliers, to prioritize fulfilling a DOD order over any commercial one. The DOD estimates that it issues approximately 300,000 DPA Title I-rated contracts annually.[72] The DOD can also request that DPA Title I ratings be applied to foreign sources, contingent upon local laws and the execution of a "Security of Supply Agreement" with the host government.[73] Theoretically, the flowdown of DPA Title I ratings throughout the supply chain should provide the DOD with both traceability and the first "call" on any critical minerals necessary for defense procurement.

Despite the "paper" strength of DPA Title I, this authority has significant limitations in practice, all of which would hamper the DOD's ability to direct critical minerals supplies to national needs in an emergency.

For example, DPA Title I ratings only apply to U.S. companies and, by request, select U.S. allies. Any nation outside this circle—an adversary or otherwise—is under no obligation to support the DOD's needs.

As indicated by the volume of DPA Title I ratings, contracting and the incorporation of critical minerals sourcing requirements into such contracts is the most common tool for managing supply chain risk. Two of the most well-known critical minerals sourcing regulations are (1) the "specialty metals clause," the colloquialism for a 1973 rule requiring the purchase of aerospace alloys and steel from U.S. or allied sources, and (2) the "sensitive materials rule," a 2019 rule that prohibits the purchase of refractory metals and rare earth permanent magnets from China, Russia, North Korea, and Iran.[74]

Historically, defense contractors have been at odds with the metals and mining sector over these procurement rules. Broadly, the domestic metals and mining sector tends to favor rules that may drive defense spending toward their facilities, which may be more costly than foreign ones. Defense contractors, on the other hand, tend to oppose rules that may complicate subcontract management and compliance costs. Particularly when a subcontractor principally serves the commercial market, DOD-unique critical minerals sourcing rules may deter participation in defense contracts.

Whenever Congress has required the DOD to implement a new critical minerals sourcing rule, the timeline for implementation has been lengthy and subject to intense advocacy campaigns. For example, reaching a final rule on the "specialty metals clause" was the subject of ongoing regulatory and legislative advocacy for almost a decade.[75] A similar battle is currently underway over the "sensitive materials rule."[76]

These two rules—the "specialty metals clause" and the "sensitive materials rule"—also highlight the well-intentioned but often inefficient promulgation of new critical minerals sourcing mandates. For example, samarium-cobalt permanent magnets are covered under both the "specialty metals clause" and the "sensitive materials rule," but since the DOD's implementation of the newer "sensitive materials rule" is executed on a contract-by-contract basis, ensuring contract compliance is highly complex and costly. Similarly, recycling is an important source of domestic production of critical minerals, but under the "sensitive materials rule," only one of the four covered critical minerals may utilize domestic recycled feedstock.[77]

Stockpiling

Surprisingly, stockpiling is hardly mentioned in JP 4-05. The limited references include one-off statements that (1) DPA Title I can compel delivery to stockpile contracts; (2) stockpiles should exist; (3) stockpiles should be released once mobilization begins; and (4) stockpiles should be rebuilt once the mobilization period ends. No other analysis or discussion of stockpile planning or management appears.

Though this brevity is refreshing, it also highlights a significant gap between DOD doctrine and NDS planning. As previously noted, NDS planning is driven by a short-war requirement, plus a long-term reconstitution phase. However, DOD doctrine calls for rapid in-crisis stockpile releases, with an indeterminate reconstitution phase. Put another way, the NDS stockpile sizing construct focuses on replacement once the fight is over, while Joint Staff doctrine wants to buy time during the emergency until other industrial base expansion programs come online.

Although the requirements generation process for stockpiling and overall doctrine have remained static, the underlying Stockpiling Act has not. The DOD requested significant reforms to this law for the FY 2023 National Defense Authorization Act (NDAA), and Congress has proceeded to implement these changes across the FY 2023 and FY 2024 NDAAs. Among other elements, this reform aims to infuse private sector best practices into stockpile management while removing statutory barriers to more efficient government operations by:

- consolidating multiple DOD critical minerals policy oversight boards into a new "Strategic and Critical Materials Board of Directors"
- authorizing multiyear procurements and general acquisition of shortfall materials
- authorizing off-take agreements from DPA Title III industrial base investment projects
- supporting feasibility studies for new critical minerals projects
- expanding the scope of potential NDS research project applicants to include U.S. allies[78]

Prior to these changes, the Stockpiling Act had remained largely untouched since 1979, and the DOD is only beginning to implement many of these reforms.[79] On the other hand, though the NDS is authorized to carry out these new functions, new funding has slowed to a trickle. After a significant one-time appropriation of $125.0 million in FY 2022 and $93.5 million in FY 2023, FY 2024 funding collapsed to only $7.6 million. These funding increments are wholly insufficient to meet the shortfall requirements for defense ($2.41 billion) and essential civilian needs ($12.21 billion) in a national emergency.[80]

Industrial Base Investment Programs

Pivoting to industrial base investment, the DOD offers an array of programs to foster the development of new critical minerals production technologies, sources of supply, and end use items for the military services. Among these are Small Business Innovation Research (SBIR) programs, basic research and qualification projects funded by the NDS, pilot or prototype demonstration projects funded by the Industrial Base Fund (i.e., the Innovation Capability and Modernization (ICAM) program), feasibility or commercial scaling projects funded by the Defense Production Act Fund, and various military service–specific organic industrial base funds.[81]

Historically, industry has expressed its frustration with the DOD's inability to bridge the "Valley of Death," where the DOD supports early-stage development of an innovative technology or product but cannot transition it to procurement by a program of record. This frustration has also been directed toward the aforementioned industrial base investment programs. However, the DOD has several recent case studies in the critical minerals sector that provide reason for optimism, with multiple companies successfully transitioning from early-stage research to commercial-scale production with or through the DOD (see Table 2).

In alignment with JP 4-05, the DOD also has awarded defense industrial base investment funds to Canadian companies, who have been considered a "domestic

Table 2: Critical Minerals Transition Programs

Company	Development Program	Scaling Program
Rare Earth Salts Separations & Refining LLC	• DLA: ($8.4 million) Rapid Innovation Fund demonstration • DLA: ($0.2 million) basic R&D study	• DPA Title III: ($4.2 million) terbium recycling program
MP Materials Corp.	• ICAM: ($0.6 million) heavy rare earth demonstration	• ICAM: ($35.0 million) heavy rare earth scaling
Lynas USA, LLC	• ICAM: ($0.6 million) heavy rare earth demonstration	• ICAM: ($258.2 million) heavy rare earth scaling
Noveon Magnetics Inc.	• DLA/SBIR: ($1.0 million) magnet recycling and production demonstration • DLA/SBIR: ($1.6 million) qualifying magnets for Excalibur, Peregrine, JDAM, and Small Diameter Bomb	• DPA Title III: ($0.8 million) magnet inventory demonstration • DPA Title III: ($28.8 million) magnet production
Graphite One (Alaska) Inc.	• DPA Title III: ($37.3 million) feasibility study	• Department of Energy, Loan Program Office: ($201 million) direct loan application
Perpetua Resources Idaho Inc.	• DPA Title III: ($59.2 million) feasibility study • Army/DLA: ($15.7 million) qualification study	• Export-Import Bank: ($1.8 billion) direct loan letter of interest
Talon Nickel (USA) LLC	• DPA Title III: ($20.6 million) nickel resource development	• Department of Energy, Manufacturing and Energy Supply Chains: ($114.8 million) nickel processing

Source: Author's analysis of awards posted at FPDS-NG and USASpending.gov and press releases by the DOD and company awardees.

source" since 1992.[82] Additionally, Congress amended the DPA to expand the scope of eligible foreign allies to include the United Kingdom and Australia.[83] Though a handful of Canadian firms have received DPA Title III awards, the legislative change for the United Kingdom and Australia is sufficiently recent that no such companies have received a DPA Title III award to date.[84]

Notwithstanding this apparent success in supporting critical minerals development through the Trump and Biden administrations, 70 percent of DOD funding for critical minerals projects—$778 million of $1.1 billion—has been derived from supplemental appropriations.[85] In other words, Congress is the principal driver behind the DOD's investments in critical minerals, not the DOD's bottom-up requirements generation and budgetary process. To that end, recent DOD budget requests suggest that critical minerals investment funding will fall to approximately $30 million per year.[86] Based on recent DPA Title III awards for critical minerals, this level of funding is sufficient to execute perhaps one or two "feasibility study" projects per year.

Waivers or Expediency Under Other Domestic Laws

With respect to other authorities to waive domestic laws or otherwise expedite critical minerals projects, the DOD does not appear to have pursued or received authorization under extant pathways for regulatory relief in U.S. environmental laws. These include, for example, national security or paramount interest pathways under the Endangered Species Act and the Clean Air Act.[87] However, additional information on DOD recommendations regarding U.S. environmental laws may be forthcoming through the FY 2025 NDAA.[88] Namely, the U.S. House of Representatives included a requirement for the DOD to report on the impact of the National Environmental Policy Act on the largest defense industrial base projects.

Environmental regulation aside, the DOD also does not appear to have actively pursued or promoted other nonregulatory pathways to streamline permitting activities for its projects—critical minerals or otherwise. Of note, only two DOD industrial base investment projects are included on the FAST-41 Covered Projects Dashboard: Perpetua Resources' antimony project in Idaho and South32's zinc-manganese project in Arizona.[89] Given the limited dataset, it is not possible to determine whether inclusion on the FAST-41 dashboard provides a meaningful benefit to project development or whether other factors—such as a U.S. government award from the DOD or another agency—are more decisive.

RECOMMENDATIONS

First, the Joint Staff and DOD critical minerals programs need to update their war planning assumptions. Senior DOD leadership, civilian and military, has clearly stated that the DOD must begin to prepare for a protracted conflict, but this view has not been reflected in the warfighting scenarios that the NDS uses for requirements generation.[90] Without needed updates to war planning, DOD base budget requests will continue to grossly underestimate critical minerals needs. Therefore, the Joint Staff should develop a war-planning scenario suitable for NDS planning to reflect DOD policy and generate more realistic estimates of defense requirements for critical minerals.

Second, the Joint Staff and DOD industrial investment and stockpiling programs should realign doctrine and program execution. The industrial base management sections of the mobilization doctrine generated by the Joint Chiefs of Staff, JP 4-05, have not changed since 1995. The document does not reflect lessons learned from (a) industrial base expansion efforts to respond to the Covid-19 pandemic or provide military assistance to Ukraine, (b) related medical or war reserve inventory distribution challenges, or (c) the management of DPA Title I allocations of scarce materials to the domestic market.[91] Moreover, the objectives established in the current doctrine (i.e., provide in-crisis response) are not matched by the NDS requirements generation process (i.e., provide for reconstitution of forces). Therefore, the Joint Staff and civilian components of the DOD responsible for industrial mobilization should update JP 4-05 or develop new doctrine to reflect how the department is likely to respond to a mobilization event.

Third, the DOD should stabilize funding for critical minerals in the base budget. The DOD has made significant progress in supporting the upstream supply chain across numerous minerals. Though these efforts only began in earnest in 2019, DOD prime contractors and major subcontractors are already integrating new domestic sources into DOD programs of record.[92] However, most of this success is being carried by one-off supplemental appropriations acts, which do not provide predictability to industry or the DOD for investment planning. Therefore, the congressional defense committees should continue to provide discrete program increases or "functional transfers" for critical minerals projects within industrial mobilization programs, such as the Defense Production Act Fund, the Industrial Base Fund, and the National Defense Stockpile Transaction Fund.

Fourth, the DOD should streamline critical minerals sourcing rules. Given rising concerns related to the United States' reliance on adversarial sources, Congress continues to legislate mandates for the DOD to restrict sources of supply for critical minerals and end-use items containing critical minerals. In some cases, the same mineral is covered under multiple sourcing rules simultaneously, with nonsensical exception structures.[93] This constantly shifting regulatory regime places a significant cost burden on all tiers of the defense industrial base, with the compliance burden especially acute at the prime contract level, given that noncompliance occurs many tiers removed from the prime contractor.

Therefore, the DOD's Office of Defense Pricing, Contracting, and Acquisition Policy should undertake an acquisition reform study focused on critical minerals sourcing. At a minimum, this study should identify the extant Defense Federal Acquisition Regulation Supplement rules and their underlying legislation for critical minerals products, describe these rules' use and exception structures, and then develop a streamlining legislative proposal for Congress. As appropriate, this proposal also should include requests for funding to support the development of military specifications, standards, or other industry-led initiatives to validate sub-tier supplier compliance.

CONCLUSION

As explored in this chapter, the central question regarding the "criticality" of a critical mineral to national defense is whether the DOD finds a classified shortfall in defense or essential civilian industry needs in a postulated wartime scenario. However simple that question may be, the answer is highly susceptible to subjective policy judgments, which flow directly from the DOD's National Defense Strategy.

Over the past seven decades, U.S. defense policy has trended toward a more optimistic appraisal of the availability of foreign sources and the severity of a conflict involving the U.S. Armed Forces. This pendulum is now swinging in the opposite direction, with a greater focus on protracted conflict.

DOD planning and posture are beginning to change for the better, particularly for rare earth elements and battery minerals. However, the preponderance of the DOD's efforts is funded by out-of-cycle supplemental appropriations acts. Critical minerals have not yet become a mainstay of the department's base budget, nor has DOD doctrine and program execution materially evolved from its immediate post–Cold War posture.

On balance, the DOD and defense industry have notched major accomplishments to secure their supply chain for critical minerals. Fully addressing this challenge, though, is a marathon, not a sprint, and the work of the DOD, Congress, and industry in this realm has only just begun.

CHAPTER 4

Driving Innovation

Critical Minerals and the Automotive Industry

By Duncan Wood and Alexandra Helfgott

ake1150 via Adobe Stock

The single biggest determinant for key minerals such as lithium, graphite, cobalt, nickel, and manganese is consumer demand for EVs themselves.

The development of the electric vehicle (EV) industry has been the single biggest driver of critical minerals demand growth—a trend that is expected to continue for years to come. This demand growth has been fueled by government incentives at both the national and subnational levels globally. However, given the inherently international nature of the EV supply chain, U.S. EV manufacturers are concerned about disruptions stemming from rising geopolitical tensions. The EV industry—which has mobilized significant investment and created nearly 100,000 jobs in recent years—will require uninterrupted access to the materials needed to produce batteries and motors.

As the Trump administration takes office, the future of the EV industry is clouded by uncertainty, with serious questions regarding the outlook for existing consumer and production incentives. The proximity of Tesla CEO Elon Musk to U.S. president Donald Trump may influence this decision, but more important will be the rationale for continued support in the context of economic security, strategic competition with China, and U.S. jobs. It is clear that U.S. automakers are already committed to the transition to EVs and hybrid vehicles, having invested billions of dollars over the past four years in building gigafactories across the United States and the rest of North America.

Uncertainty for the sector also stems from shifts in consumer preferences and interest rates, technological advancements in battery chemistries, and a slower-than-expected expansion of the charging network. Moreover, the uncertainty currently affecting the EV industry in the United States has a knock-on effect on the global market for critical minerals, particularly regarding U.S. and allied countries' investments in the critical minerals supply chain.

This chapter will examine the drivers of growing demand for EVs, the knock-on effects on demand for critical minerals, and the challenges facing the supply chain. It will also highlight the importance of innovation in the EV battery sector to reduce the industry's vulnerability to interruptions and shortcomings in the critical minerals supply chain. A combined approach of reducing demand for critical minerals through innovation, boosting domestic supply of those minerals, and working closely with allies to secure U.S. supply chains will provide certainty and stability for the market, protecting U.S. investments, jobs, and competitiveness.

THE IMPORTANCE OF THE AUTOMOTIVE MARKET IN THE UNITED STATES AND NORTH AMERICA

The auto industry is a cornerstone of North American trade, accounting for 22 percent of total trade under the United States-Mexico-Canada Agreement (USMCA).[94] It supports a staggering 42.2 million jobs across the region, both directly and indirectly.[95] In the United States, the industry directly employs 9.7 million people and supports an additional 11 million indirect jobs, highlighting its critical role in the nation's economy.[96] In Mexico, the sector provides 1 million direct jobs and contributes to 20 million indirect positions, underscoring its importance in driving economic development.[97] Canada, while smaller in scale, still benefits significantly, with 500,000 jobs tied to the industry.[98] This interconnected workforce demonstrates the auto sector's immense economic impact and its role as a vital driver of prosperity across North America.

In 2023, North America produced approximately 3.6 million EVs, and the industry created over 200,000 direct EV-related jobs across manufacturing, battery production, and infrastructure development.[99] It is important to note that much of this production relies on cross-border, tariff-free trade under the USMCA.

Growing Demand

The single biggest determinant of demand for key minerals such as lithium, graphite, cobalt, nickel, and manganese is consumer demand for EVs themselves. This demand has grown substantially in recent years, with EV sales reaching nearly 14 million cars globally in 2023.[100] The International Energy Agency (IEA) estimates that global demand for critical minerals driven by EV production was under 2 million metric tons in 2020 but is projected to exceed 30 million metric tons by 2030, representing approximately 75 percent of the minerals required for clean technology (cleantech). By 2050, EV demand alone is expected to account for over 130 million metric tons, or roughly 90 percent of the total mineral demand for cleantech. However, this growth is mostly concentrated in the United States, Europe, and China—countries where the use of personal vehicles is more common—though EV sales in Southeast Asia and Brazil are picking up speed, largely due to government subsidies and the availability of low-priced Chinese EVs.[101]

In the United States, major decisions by auto companies to produce more EVs, or even shift to 100 percent EV production, have begun to fundamentally alter the market. General Motors' 2021 announcement of plans to transition to 100 percent EV production by 2035 marked a watershed moment for the domestic industry.[102] President Joe Biden's goal of having 50 percent of auto sales be electric by 2030 was another signal to the market that demand will grow significantly.[103] However, the major factor driving the EV market in recent years has been the Inflation Reduction Act (IRA).

The IRA, which is discussed in detail in Chapter 6, provided important tax credits for new EV sales. Section 30D of the IRA includes a $7,500 tax credit specifically for light-duty EVs for individuals earning less than $150,000 or families earning less than $300,000.[104] Half of the tax credit is allocated for batteries manufactured with materials mined in the United States or countries with which the United States has a free trade agreement. Notably, materials recycled within North America are also eligible. The percentage of the value of critical minerals mined or processed in these countries will increase from 40 percent in 2023 to 80 percent by 2027.[105]

Starting in 2025, batteries utilizing critical minerals mined or processed by foreign entities of concern (primarily referring to Chinese-owned firms) will not be eligible for the tax credit. The other portion of the EV credit applies to vehicles with batteries manufactured or assembled in North America, with the percentage of components manufactured in the region set to increase over time—from 50 percent in 2023 to 100 percent by 2029.[106] The inclusion of this requirement in the IRA underscores the region's importance in the auto industry and emphasizes the critical role of regional integration in strengthening the industry.

Recent Trends

Since the passage of the IRA, the United States has seen considerable growth in EV sales, reaching 7.9 percent of total sales in 2023.[107] Driven largely by these incentives, sales of EVs have risen significantly in recent years in the United States. Despite a slow year in 2023, sales of EVs in the United States have rebounded rapidly in 2024. In the third quarter of 2024, EV sales increased by 11 percent year-on-year, while EVs as a share of all automotive sales in Q3 reached 8.9 percent.[108] The EV industry will soon account for a tenth of all auto sales in the United States.

However, progress has been far from linear. According to the U.S. Energy Information Administration, sales of hybrid, plug-in hybrid, and battery electric vehicles (BEVs) grew to over 16 percent of total new light-duty vehicle sales in the United States in 2023.[109] In the first part of 2024, however, EV sales declined, with nearly a 1 percent decrease in hybrids, plug-in hybrids, and BEVs sold in the first quarter of 2024 compared to the fourth quarter of the previous year.[110] This decrease in demand for EVs in 2023 and 2024 has pushed automakers to rethink their strategies, particularly as hybrid vehicles gain more traction.

Despite the fact that sales of EVs have risen significantly in the United States, these figures are disappointing compared to global numbers. In China, BEVs are projected to account for 50 percent of all light vehicle sales by the end of 2024—an impressive total with far-reaching implications for the global

Figure 1: Quarterly U.S. Light-Duty Vehicle (LDV) Sales by Powertrain, January 2014–June 2024

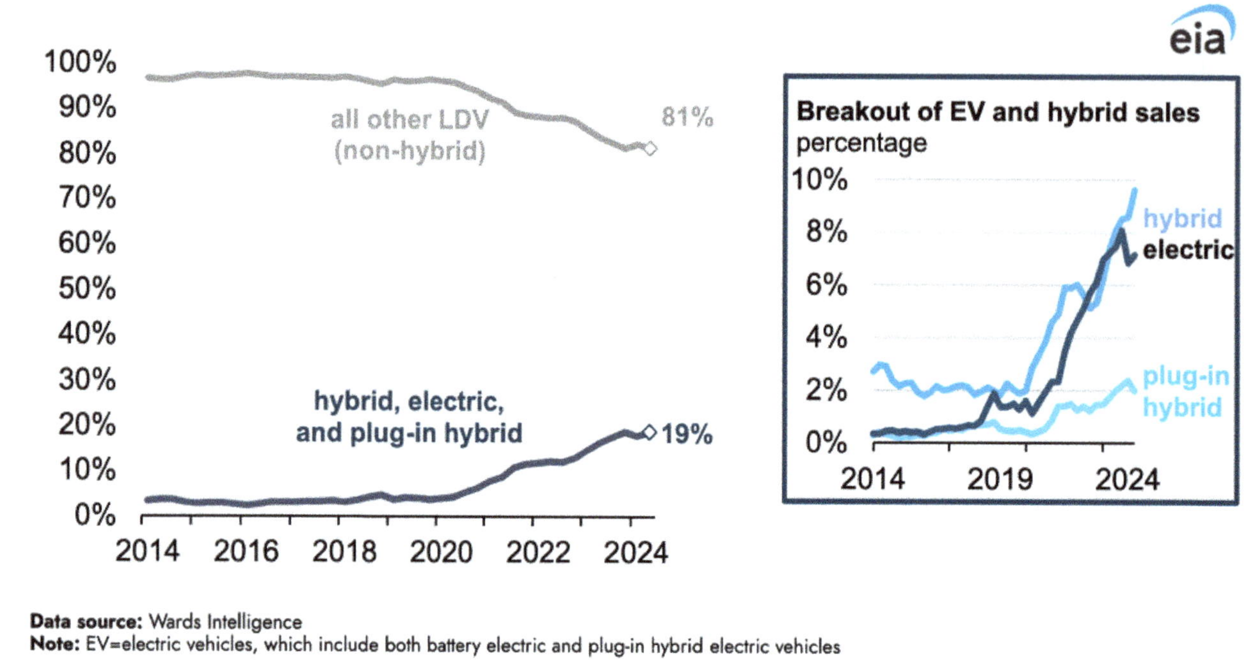

Source: Monica Abboud, "U.S. share of electric and hybrid vehicle sales increased in the second quarter of 2024," U.S. Energy Information Administration, August 26, 2024, https://www.eia.gov/todayinenergy/detail.php?id=62924.

oil and critical minerals markets. According to the Alliance for Automotive Innovation, Chinese EV manufacturing is comparable to the entire output of the U.S. auto industry.[111]

The story of EV sales in Europe is less straightforward. After rapid growth in 2021 and 2022, 2023 saw a slowdown in global EV sales due to rising inflation and the end of government subsidies, particularly in Europe. This was most notable in Germany, where sales dropped by 37 percent in July 2024 following the government's termination of EV subsidies.[112] Registrations for hybrid vehicles in Europe reached 24 percent in July 2024, while EV registrations were just 13.6 percent—nearly a full percentage point lower than during the same time period the previous year.[113] The slowdown in Europe continued throughout the first half of 2024, but EV sales saw an increase in the third quarter.

What these statistics show is that linear development of the EV industry is not guaranteed. Although numbers continue to rise, much work remains to drive consumer demand and ensure the industry's sustainability in the United States. While long-term growth is expected, it is important for investors and policymakers to understand that progress will involve leaps forward and some steps back before a more widespread shift to EVs occurs.

THE EV SUPPLY CHAIN

The first phase of the EV battery supply chain, often referred to as the upstream portion of the supply chain, is raw mineral extraction. This phase is arguably one of the most important, as it forms the fundamental basis of the EV battery. The list of critical minerals essential for EV battery manufacturing is extensive and includes manganese, graphite (and graphene), lithium, nickel, and cobalt, among others.

The midstream phase of the EV battery supply chain entails the processing and refining of raw materials. This typically requires high-heat or chemical-based treatments to transform the raw materials into what will eventually be cathode and anode active battery materials. Rare earth metals—a group of 17 elements—are used in various clean energy technologies for their "permanent magnetic properties."[114] In the specific case of EVs, they are primarily used in magnets for EV motors and as catalysts for battery fuel cells. Critical minerals processing tends to be regionally concentrated, and it is more common than not for extraction and refining to occur at separate facilities.

The third stage in the EV battery supply chain, the downstream phase, involves assembling battery cells into modules, which include battery management systems, electronics, and sensors. These modules are then packaged and sold to automakers, although some manufacturers produce and install their own battery packs.

The final process in the EV battery supply chain is reuse and recycling. Reuse entails "disassembly of the pack, testing module/cells, and repackaging."[115] Pyrometallurgy, hydrometallurgy, and direct recycling are currently the three most viable options for lithium-ion battery recycling, though new technological advances are emerging. Both reuse and recycling are logistically challenging—not only from an economic and regulatory standpoint but also in terms of the basic logistics of transporting the batteries. Moreover, the manufacturer-specific nature of the batteries adds to the cost of recycling.

EV DEMAND AND THE CRITICAL MINERALS SUPPLY CHAIN

It is abundantly clear that the shift from traditional internal combustion engine (ICE) vehicles to EVs will require significantly increased quantities of critical minerals for EV production. Whereas a traditional ICE vehicle uses an estimated 34 kg of critical minerals, an EV uses approximately 200 kg, primarily for the electric motor and battery.[116] However, this is likely to change over time as battery size, battery chemistry, modularity, and consumer preferences continue to change. While larger battery sizes may increase mineral demand, shifts in battery chemistry will impact the mix and proportions of metals required in those batteries.

The IEA makes the important point that the percentage of electric sport utility vehicles (SUVs) being sold has a significant impact on the critical minerals supply chain:

> Larger electric car models have a significant impact on battery supply chains and critical mineral demand. In 2023, the sales-weighted average battery electric SUV in Europe had a battery almost twice as large as the one in the average small electric car, with a proportionate impact on critical mineral needs. . . . [I]f all electric SUVs sold in 2023 had instead been medium-sized cars, around 60 GWh of battery equivalent could have been avoided globally, with limited impact on range. Accounting for the different chemistries used in China, Europe, and the United States, this would be equivalent to almost 6,000 tonnes of lithium, 30,000 tonnes of nickel, almost 7,000 tonnes of cobalt, and over 8,000 tonnes of manganese.[117]

According to the IEA, copper, cobalt, nickel, lithium, rare earth elements, and aluminum are the minerals in highest demand.[118] Demand for these critical minerals is projected to grow fourfold under the IEA's Sustainable Development Scenario by 2040.[119] However, this projection is subject to three key external factors: evolving technology, the development and implementation of governments' clean energy policies, and the demand for EVs.[120] All three factors are currently experiencing a high degree of uncertainty.

The two-way connection between the EV market and critical minerals prices is exemplified by the recent collapse in lithium prices, which has impacted the competitiveness of EVs in automotive markets. After peaking at over $79,637 per ton in December 2022—driven by soaring demand for EVs—lithium prices fell to less than $11,000 per ton by September 2024.[121] This decline was caused by several factors, including high interest rates, a weak Chinese economy, and market manipulation. Additionally, new resources coming into production globally and lower-than-expected EV sales following the initial surge in the United States, China, and Europe were significant contributors. Stagnating EV sales in 2023 had a profound and rapid impact on the lithium market. For instance, earlier in 2024, Ganfeng Lithium Group reported a net loss of $107 million and announced plans to limit capacity expansions as a "glut of supply overwhelmed slower-than-expected demand growth from electric-vehicle makers, driving spot carbonate prices to a three-year low."[122]

Some experts predict that the current oversupply of lithium will last until 2027.[123] In the interim, there is hope that new extraction—and, more importantly, processing—projects will come into operation around the world, particularly in the United States and U.S.-friendly countries.

The difficulty in estimating demand projections is further exacerbated by governments' evolving approaches to clean energy policy development and implementation. For example, changes in administrations in the United States and a potential move away from the incentives outlined in the IRA will play a significant role in determining short-term demand.

Other factors influencing future EV demand include local- and state-level incentives and regulations. Just as the IRA's clean vehicle tax credits boosted demand for EVs after 2022, California's rapid shift toward an EV-friendly regulatory framework and higher gasoline prices had a similar effect. While California has the best-known incentives, many other states have followed suit.[124]

Consumer financing innovations also have the potential to incentivize higher EV sales and end-of-life recycling. In the United States, the IRA succeeded in driving new financing from auto firms such as Hyundai, whose vehicles were not eligible for IRA tax credits. In response, Hyundai implemented its own financing mechanisms to match the value of these credits. At the industry level, there is room for even greater innovation. For example, a recent paper on the cobalt supply chain proposed a "lease-to-recycle" model for batteries to drive EV adoption and enhance the potential for recycling battery metals.[125]

Private sector funding for EVs and battery facilities has significantly shaped the industry's development in the United States. Between 2018 and 2024, the private sector has announced investments of $90 billion in battery facilities and $33 billion in EV facilities.[126] Notably, states with Republican governors have been more successful in securing this funding, with the southern region of the United States receiving over $68 billion in regional investments. In comparison, the Northeast received just $300 million, while the West Coast garnered $13.3 billion—well below the South's total.[127]

One additional factor influencing the demand for critical minerals for EVs is the buildout of the charging network. Persistent consumer concerns about locating charging stations, charging speeds, and the maintenance and distribution of these stations are often grouped under the term "range anxiety." However, as EV ranges improve with advances in battery technologies, it may be more accurate to discuss charging convenience. Potential EV buyers in urban areas face unique challenges regarding access to charging infrastructure. For example, availability in their neighborhoods—or, more specifically, in apartment buildings—may be limited. Drivers who lack garages and rely on street parking face even greater concerns about convenience.

The Biden administration's National Electric Vehicle Infrastructure program allocated $7.5 billion to the construction of 30,000 charging ports across the United States, with a particular focus on Alternative Fuel Corridors. These funds are intended to be distributed through state governments. However, as of November 12, 2024, only 102 charging ports at 25 charging stations in nine states had been opened.[128] Several factors have contributed to this slow rollout, but a significant portion of the total funding—around $4 billion—has already been committed to the states. Consequently, the buildout will accelerate over the next few years, as these funds cannot easily be rescinded by the federal government.

THE ROLE OF TECHNOLOGICAL INNOVATION

Innovation in battery chemistry and design is likely to play a crucial role in shaping the demand for critical minerals in the EV industry. While the internal combustion engine evolved slowly over the past 150 years, the EV industry is seeing rapid, profound, and unparalleled advances in battery chemistry, design, and efficiency.

Mineral demand for EV batteries depends on the cathode and anode chemistries of the batteries but is ultimately influenced by evolving technologies that have the potential to alter the mineral composition of EV batteries. For example, an NMC (nickel manganese cobalt oxide) battery uses half as much nickel as an NCA (nickel cobalt aluminum oxide) battery but requires eight times more cobalt. LFP (lithium iron phosphate) batteries, by contrast, require 50 percent more copper than NMC batteries but do not use nickel, cobalt, or manganese.[129] NMC batteries typically last around 2,000 cycles—a battery cycle is the process of a battery being fully charged and then discharged—though their capacity declines significantly after 1,000 cycles. By comparison, LFP batteries are more durable, retaining 80 percent of their capacity after 3,000 cycles. This longer lifespan and more consistent performance over time help to explain the rise of LFP batteries in China, where they now dominate the market, accounting for over 60 percent of current vehicle sales.[130] A similar trend is emerging in the United States and Europe.

A reduced emphasis on nickel and cobalt not only decreases U.S. exposure to Chinese-controlled extraction and processing. It could also transform the debate over deep sea mining, removing a key justification for the extraction of polymetallic nodules rich in cobalt and nickel from the seafloor.

Existing and emerging battery chemistries and technologies that will play a major role in the near future include the following:

Existing Technologies

a. **Lithium-Ion Batteries (Li-ion):** The current industry standard, Li-ion batteries, offers a balance of energy density, lifespan, and cost. These batteries encompass multiple chemistries, including lithium nickel manganese cobalt oxides (NMC) and lithium nickel cobalt aluminum oxides (NCA). While Li-ion batteries continue to power most EVs on the road today, they are rapidly losing ground to LFP batteries. The continued use of Li-ion batteries sustains high demand for lithium, nickel, and cobalt.

b. **Lithium Iron Phosphate (LFP) Batteries:** Safer and cheaper than traditional Li-ion chemistries but with lower energy density, LFP batteries are rapidly challenging the status quo in EV sales. A shift to LFP means lower demand for cobalt and nickel, with increased use of more abundant iron and phosphate.[131]

Emerging Technologies

a. **Solid-State Batteries:** These batteries, though still in development, promise higher energy density and improved safety by replacing liquid electrolytes with solid materials. The development of solid-state technology will reduce the need for cobalt and nickel but continue reliance on lithium and possibly new solid electrolytes, like lithium metal.

b. **Graphene-Based Batteries:** A technology still in development, graphene batteries involve the integration of graphene into the cathode and anode to significantly improve energy density, charging speeds, and battery life. Often considered a "wonder material" due to its lightweight nature and superior performance in various applications, graphene's high conductivity allows for faster charge transfer, and its durability supports longer-lasting batteries, making it a promising material for future EV battery advancements. A shift to this technology would increase demand for graphite and graphene, the latter being essentially the building block for graphite.

c. **Sodium-Ion (Na+) Batteries:** An emerging alternative to Li-ion batteries, Na+ batteries use sodium instead of lithium as the primary charge carrier. They can be cheaper and more sustainable than lithium (due to the natural abundance of sodium), but they have lower energy density. Na+ batteries have potential applications in stationary energy storage and some EVs. Growth in their use would reduce dependence on lithium.

d. **Lithium-Sulfur (Li-S) Batteries:** This emerging subset of lithium-ion chemistries has the potential for very high energy density but faces issues with durability and lifespan. Li-S batteries have a higher energy density than traditional Li-ion batteries and rely on sulfur, which is abundant and inexpensive, instead of nickel and cobalt.[132]

Battery design is also an important factor. Currently, EV batteries are relatively standardized across the industry. However, experts anticipate that future developments will introduce modular designs, enabling customers to tailor their batteries to meet specific needs, such as prioritizing extended range or enhanced performance.[133]

POLICY RECOMMENDATIONS

The return of the Trump administration is expected to bring major changes to U.S. EV policies, potentially undoing or slowing down key Biden-era efforts. Trump has expressed skepticism toward EVs and may push to roll back regulations and eliminate the $7,500 federal tax credit for EV buyers. Such moves could create hurdles for automakers and dampen consumer interest in EVs, raising concerns across the industry.

Funding for EV infrastructure, including nationwide charging stations, may also face cuts or redirection, while emissions standards could be relaxed. Despite these federal policy shifts, states like California are expected to continue pursuing aggressive climate and EV initiatives. This may lead to legal battles reminiscent of previous conflicts over state versus federal authority on environmental regulations.

Three key recommendations are proposed to address the challenges posed by potential policy shifts:

1. **Prioritize investing in innovation.** This can be done through two avenues. First, the Department of Energy should fund research into battery technologies that use smaller quantities of critical materials, explore applications for more abundant resources (such as sodium and graphene), and reduce both costs and dependency on foreign suppliers. If U.S. innovation can be harnessed effectively to reduce the vulnerabilities of the EV battery supply chain, jobs, investments, and U.S. competitiveness can be protected and promoted. Second, the government should fund innovation to strengthen the circular economy. The United States needs to advance recycling technologies and policies to maximize resource efficiency and minimize waste in battery production and use. Battery metals that have already been processed to the point where they can be used in batteries require less processing when recycled than newly extracted metals. The Department of Energy must continue to fund research into battery design to enhance the potential for recycling battery metals.

2. **Use tariffs judiciously and promote stable tariff policies.** In November 2024, Trump announced that his administration would impose a 25 percent tariff on all goods entering the United States from Mexico and Canada (and an additional 10 percent on goods entering from China).[134] This would be highly disruptive for the auto industry, which has a deeply integrated mine-to-market supply chain in North America, given that mining, processing, and manufacturing occur across the continent. For example, Canada is the biggest supplier of nickel alloys that are vital for EVs. The United States has only one operating nickel mine—Lundin's Eagle Mine in Northern Michigan. However, since the United States does not have a single completed nickel refinery, the entirety of nickel output from Lundin's Eagle Mine is exported to a refinery in Sudbury, Canada, and the refined ore is then sold back to U.S. firms for manufacturing. Given the importance of the automotive sector to the United States, the Trump administration should minimize tariffs with allied nations. Importantly, policies also need to be stable—when mining, processing, and manufacturing firms lack certainty, they are more likely to withhold investment in the supply chain.

3. **Expand production tax credits for all domestically manufactured vehicles.** The IRA has shown that industrial policy can drive huge levels of investment and meaningful job growth. It is time for the U.S. government to provide such support for the entire U.S. auto industry, privileging U.S. and North American content in ways that help the industry compete against Chinese competition, both domestically and abroad. IRA credits for EV production have driven hundreds of billions of dollars in investment in the United States—across traditionally red and blue states. Providing similar support for the entire auto industry would likely drive investment from U.S. and foreign auto manufacturers and bring manufacturing job growth to the United States. However, it is important to note that the EV industry warrants larger production tax credits due to its more nascent technology, as well as that Chinese EVs have become dominant in many overseas markets. Production tax credits will allow the U.S. EV industry to develop and achieve a larger global footprint.

CONCLUSION

Today's EV critical minerals supply chains are based on yesterday's estimated demand for existing battery chemistries, and as this chapter has shown, there is a rapid evolution in battery technologies underway. Whereas the internal combustion engine remained largely unchanged throughout most of its existence, battery technology is progressing at a breakneck pace. Impressive innovation already underway in the EV battery industry will continue to shift critical minerals demand, and that necessitates a more liberal approach to the U.S. supply chain strategy. The United States must take advantage of its incentive-based system that prioritizes public-private cooperation to foster innovation and not only match Chinese advances but actually surpass them.

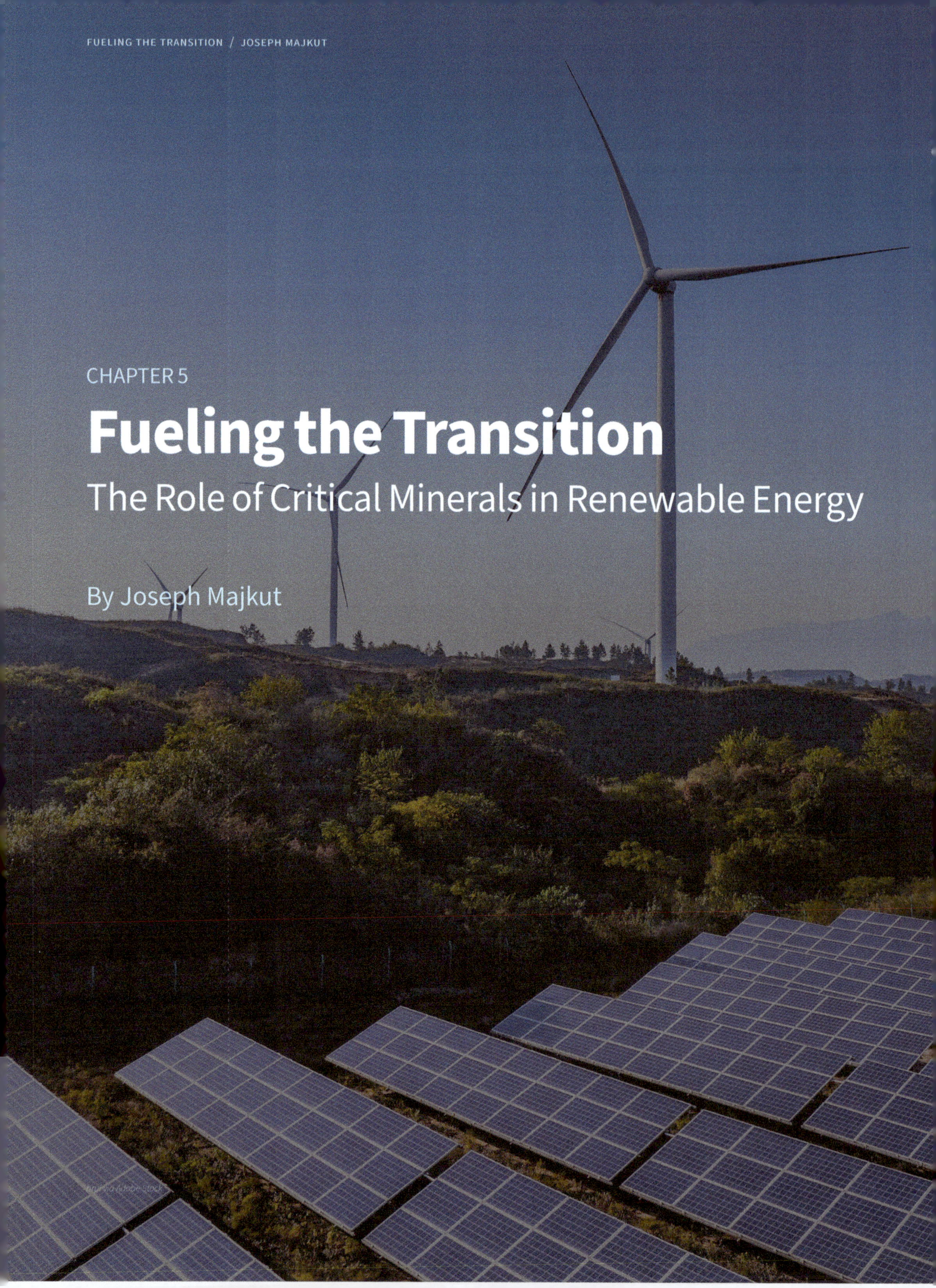

CHAPTER 5

Fueling the Transition

The Role of Critical Minerals in Renewable Energy

By Joseph Majkut

The renewable energy sector is increasingly central to both economic growth and national security, making [such] supply chain vulnerabilities untenable.

Once a niche source of electricity generation, renewables are rapidly increasing their share in the global power system. Led by solar, investments in renewable energy compromise more than 90 percent of power sector investment.[135] Renewable sources accounted for 30 percent of total global power generation in 2023 and have been growing faster than the power sector overall.[136] This has been driven in part by a decline in costs: Between 2009 and 2019, the price of wind power dropped by 70 percent and solar by 89 percent, making both technologies competitive with other new sources of electricity and an attractive economic option.[137]

In the United States, renewables are already a significant contributor to the grid and have a promising outlook. With the exception of hydropower, renewable sources contributed 15.7 percent of U.S. power generation in 2023, with wind providing 10.2 percent and solar 3.9 percent.[138] The tax credits in the Inflation Reduction Act (IRA) are further improving the outlook for both deployment and domestic manufacturing of renewables components. Projections suggest that by 2050, wind and solar could supply anywhere from 44 to 85 percent of U.S. power generation.[139]

Such rapid growth in renewables, however, brings its own set of challenges. Building renewable energy infrastructure, especially wind turbines and solar photovoltaic (PV) panels, requires substantial inputs of both common materials such as concrete, steel, and aluminum and specialized minerals such as cobalt, tellurium, and rare earth elements (REEs). Wind projects incorporate 18 of the 50 critical minerals identified by the U.S. Geological Survey (USGS) in its 2022 Critical Minerals Assessment, while solar projects require 15 of them.[140] As demand for renewables grows, so too will demand for those critical inputs. And as manufacturing of solar PV and wind turbine components is reshored, the economic stakes of supply disruptions to critical minerals will increase.

The renewable energy sector is increasingly central to both economic growth and national security, making such supply chain vulnerabilities untenable. To secure the future of renewable energy and meet ambitious climate targets, U.S. policymakers should address these vulnerabilities as part of a comprehensive approach to building a larger and more diversified supply chain for clean energy. This chapter focuses on two materials that already have the attention of policymakers and are key to deploying the most used technologies: REEs for wind turbines and polysilicon for solar PV. As such, they are also key to U.S. goals related to derisking supply chains, ensuring economic security, and furnishing the energy transition.

MATERIAL NEEDS

The International Energy Agency (IEA) estimates that global demand for minerals used in wind and solar PV could increase by a factor of about three in scenarios involving rapid emissions reductions by 2040. As a challenge for markets and policymakers to meet, this one appears "manageable." By comparison, the demand for battery minerals is expected to grow more dramatically, with EVs and grid-scale batteries driving up demand for lithium by over 40 times today's production levels.[141]

This is in line with other global assessments from academic literature, which suggests that achieving ambitious climate targets through renewables deployment will require significant increases in key minerals. Stringent climate policies could triple the demand for certain REEs used in wind turbines, while the demand for polysilicon in solar PV could more than double. Though such rapid growth would surpass historical trends, global reserves are expected to be sufficient to meet cumulative demand, especially as recycling of older turbines and solar PV cells becomes economical.[142]

China holds a dominant role as a producer, refiner, and processor of both REEs and polysilicon. These materials have already faced export controls and tariffs, as well as prohibitions on imports into the United States under the Uyghur Forced Labor Prevention Act.[143] These disruptions—and their economic and social consequences—are motivating U.S. efforts to secure stronger supply chains of these materials.

Wind

Over the past 10 years, the United States has installed an average of 10 gigawatts (GW) of wind capacity per year, resulting in nearly 150 GW of total installed wind capacity by the end of 2024. Energy modeling suggests that with the renewable tax credits from the IRA, wind capacity deployments could increase to 15 or 20 GW per year, which would double mineral demand over the next decade.[144] However, in its assessment of the material requirements for renewables, the National Renewable Energy Lab (NREL) projected that achieving ambitious climate targets could require wind capacity deployment of up to 90 GW per year.[145]

As wind energy deployment grows, the demand for minerals increases proportionally for each input—with REEs standing out as a particular area of vulnerability. These elements are key to the efficient and lightweight permanent magnets that allow for large wind turbines. The U.S. wind industry already consumes a small fraction of global production (see Table 1), and further expansion would require a substantial portion of today's global output. The huge relative share of global production that will be required to serve U.S. demand alone indicates there is significant transition risk associated with REE supply, a judgment echoed in the U.S. Department of Energy (DOE)'s 2023 Critical Minerals Assessment.[146] Expanded wind deployment globally will require much higher production of REEs, with the IEA estimating that ambitious climate targets would require REE production to increase by up to a factor of 7 by 2040.[147]

In addition to future availability, supply concentration is a particular concern for REEs. China accounts for about 60 percent of REE extraction, 85 percent of processing, and over 90 percent of permanent magnet production.[148] Its dominant role in refining and processing is particularly important. The United States is the second-largest producer of REEs globally, behind China, but U.S.-produced rare earths have historically been sent to China for processing.[149]

The critical mineral supply chain for wind power, particularly when it comes to REEs, is vulnerable and potentially problematic. Given the importance of REEs and permanent magnets throughout the economy, they are already being addressed by policymakers. The United States is home to one major rare earth mine, the Mountain Pass Rare Earth Mine and Processing Facility in California. In 2023, its operator, MP Materials, announced plans to recommission processing facilities at the site to separate and refine high-purity

Table 1: Availability of Vulnerable Wind Materials Needed to Satisfy Annual U.S. Wind Deployment

Material	Leading U.S. Import Source (2016–2019)	Current Production (millions kg/yr)	Projected Availability (millions kg/yr)	Percentage of Current Global Production Required for U.S. Wind Deployment	
				Current Levels (10 GW/yr)	**Potential Future Levels (90 GW/yr)**
Dysprosium	China	2.4	44	0.8–3%	9–28%
Neodymium	China	40.8	1,200	1–4%	10–35%
Praseodymium	China	14.4	370	<0.1%	3–7%
Terbium	China	0.5	10	<0.1%	1–2%

Source: Aubryn Cooperman et al., Renewable Energy Materials Properties Database: Summary (Golden, CO: National Renewable Energy Laboratory, 2023), 20–21, https://www.nrel.gov/docs/fy23osti/82830.pdf.

neodymium-praseodymium (NdPr) oxide, essential in permanent magnets.[150] The DOE is also supporting an MP Materials facility in Texas that will produce rare earth magnets.[151] Additional investments are being made to establish REE processing in the United States. Australian company Lynas Rare Earths, with support from the U.S. Department of Defense, is constructing a facility in Hondo, Texas, that will process both light and heavy REEs supplied by mines in Australia.[152] These efforts are designed to create a full supply chain for REEs that can meet some of U.S. demand and diversify away from China.[153]

Solar

Over the past decade, the United States has added over 100 GW of solar PV capacity, culminating in 131 GW of installed capacity by the end of 2024.[154] Unlike the relatively steady growth of wind power, solar deployment has been accelerating. Projects planned and under construction are expected to increase solar capacity by 38 percent between 2023 and 2024 alone. Looking ahead, forecast models suggest IRA incentives could help grow solar capacity by 30 to 60 GW per year by 2035.[155] Similarly, the DOE has projected high-end deployment rates of nearly 70 GW per year through 2035, tapering to 50 GW per year by 2050.[156]

In high-deployment scenarios for solar, the DOE found that while demand would increase for materials such as silver, silicon, and aluminum, these increases would be moderate compared to today's production and rising global demand.[157] In the case of silicon, the maximum consumption estimate for 2030 is 157,000 tons per year, which remains small compared to today's annual production of 8 million tons.[158] Overall, by 2030, rapid solar deployment in the United States is projected to account for about 0.6 percent of current global demand for aluminum, 2.0 percent for silicon, and 3.7 percent for silver.[159] Despite the diverse material requirements, the quantities needed for U.S. solar power make it relatively safe from supply limitations even in high-use cases.

Although silicon is abundant, solar PV systems require silicon products that are purified and refined. Most solar PV modules currently being deployed use crystalline silicon (c-Si) cells, which use polysilicon as a key input. Polysilicon is essential for the semiconducting properties that convert sunlight into energy and is less costly than monocrystalline silicon. Solar-grade polysilicon must be of very high purity and thus requires refining and processing.[160]

China dominates solar-grade polysilicon production, accounting for 85 percent of global manufacturing capacity.[161] A significant portion of that capacity is located in Xinjiang Province, raising concerns about the potential use of forced labor in the solar supply chain.[162] Additionally, the environmental impact of Chinese solar manufacturing is notable, as reliance on a coal-heavy grid results in Chinese polysilicon having higher emissions. More broadly, the high concentration of supply in China raises concerns about supply chain dependence for U.S. manufacturers of downstream products, including solar cells and modules. A recent study found that reshoring PV manufacturing to the United States would decrease the sector's emissions intensity by 30 percent.[163]

In pursuit of supply diversification, tax credits and trade restrictions on Chinese products could incentivize the relocation of polysilicon manufacturing to the United States. The DOE is tracking 69,000 tons per year of U.S. polysilicon production capacity, which compares favorably to the 157,000 tons per year by 2030 in a high-solar-deployment scenario.[164] This capacity currently stems from three facilities that have been reopened or expanded in recent years: Hemlock Semiconductor, REC Silicon, and Wacker Polysilicon North America.[165] Michigan-based Hemlock Semiconductor, for example, announced an expansion in 2022, and REC Silicon is reopening a formerly shuttered manufacturing facility in Moses Lake, Washington.[166] In addition, the DOE allocated funds in April 2024 to support Tennessee-based Highland Materials for a manufacturing plant that will produce 16,000 tons of solar-grade polysilicon.[167] These projects are being enabled by the IRA's 45X Advanced Manufacturing Production Tax Credit, under which companies can claim $3 per kilogram for manufacturing solar-grade polysilicon, as well as additional credits for manufacturing PV wafers, cells, and modules, as well as other solar-related components.[168]

RECOMMENDATIONS

Specific policy interventions have enabled the progress that U.S. industry has made in securing supply chains for REEs and polysilicon. But the job is not yet complete, and projects are still coming online. Looking forward, these programs will have to be evaluated on their ability to foster commercially viable projects that participate meaningfully in the supply chain for an expanding renewables industry. Commercial competition, especially from Chinese firms, will be intense. Policy interventions will be successful when they help build a resilient supply chain for the United States.

- **The DOE and other agencies should seek opportunities to coordinate different policy tools to develop integrated supply chains.** For REEs, the key will be to develop expertise and technology that can refine and process these minerals for multiple uses and incorporate them into permanent magnets. This will help the United States take advantage of its domestic production and bring part of the supply chain out of China. Additional projects can take advantage of the IRA's 48C tax credits and support from the DOE's Loan Program Office, which can fund critical mineral projects at all stages of production: extraction, processing, manufacturing, and recycling.[169]
- **Policymakers should cooperate with allies such as Australia and Canada in securing REEs.** Expanding collaboration through the Minerals Security Partnership or bilateral trade agreements could help source additional REEs for processing or manufacturing in the United States. In particular, expanding the Defense Productions Act's definition of "domestic source" to include Australia would grant mining and processing facilities in that mineral-rich country access to direct support and procurement authorities.[170]
- **Polysilicon projects under development should be fostered and linked to higher levels of the solar supply chain.** This will likely happen with support from the 45X tax credit, which applies to various stages of the solar manufacturing chain, including production of wafers, cells, and modules. Polysilicon projects at these stages can be encouraged to take advantage of domestic resources; wafer and cell manufacturing facilities in particular should be supported via the clean manufacturing tax credits and other programs to encourage local demand. Revising domestic content requirements for solar tax credits to reflect the upstream portions of the supply chain would further incentivize domestic polysilicon.[171] Such integrated industry is key to building ecosystems of innovation that reduce cost and build resilient supplies.
- **The new administration should address the role that tariffs and trade restrictions on Chinese products and firms will play in its strategy for securing supply chains.** In addition to enforcing trade restrictions on Chinese polysilicon, the Biden administration is increasing tariffs on rare earth magnets and Chinese-produced solar cells to defend domestic producers.[172] In addition, there are multiple congressional proposals to exclude Chinese-sourced technology and minerals from U.S. tax credits.[173] Such restrictions should be designed to address unfair trade practices, the high carbon intensity of Chinese production, and human rights but should not contribute to trade tensions through unfair protectionism. Moreover, fully excluding Chinese firms or technology—which could still be brought into the United States through joint ventures or licensing—would hamper efforts to establish resilient supply chains in a cost-effective way.

CONCLUSION

The United States is experiencing significant growth in renewables deployment, supported by climate policy and increasingly good economic outlooks for wind and solar power. While concerns over mineral supply constraints are warranted, they appear manageable in the case of the increased demand expected from renewables (though certain REEs will need attention). By contrast, concerns over China's dominance in the production and processing of REEs, other critical minerals, and polysilicon are well founded. Investments in domestic production, refining, and recycling should help mitigate the risks of supply disruptions and reduce dependence on foreign sources, setting the stage for long-term energy security.

SECTION 2

Progress Under the Biden Administration

Important but Incomplete

CHAPTER 6

An Evaluation of the Inflation Reduction Act

By Gracelin Baskaran and Meredith Schwartz

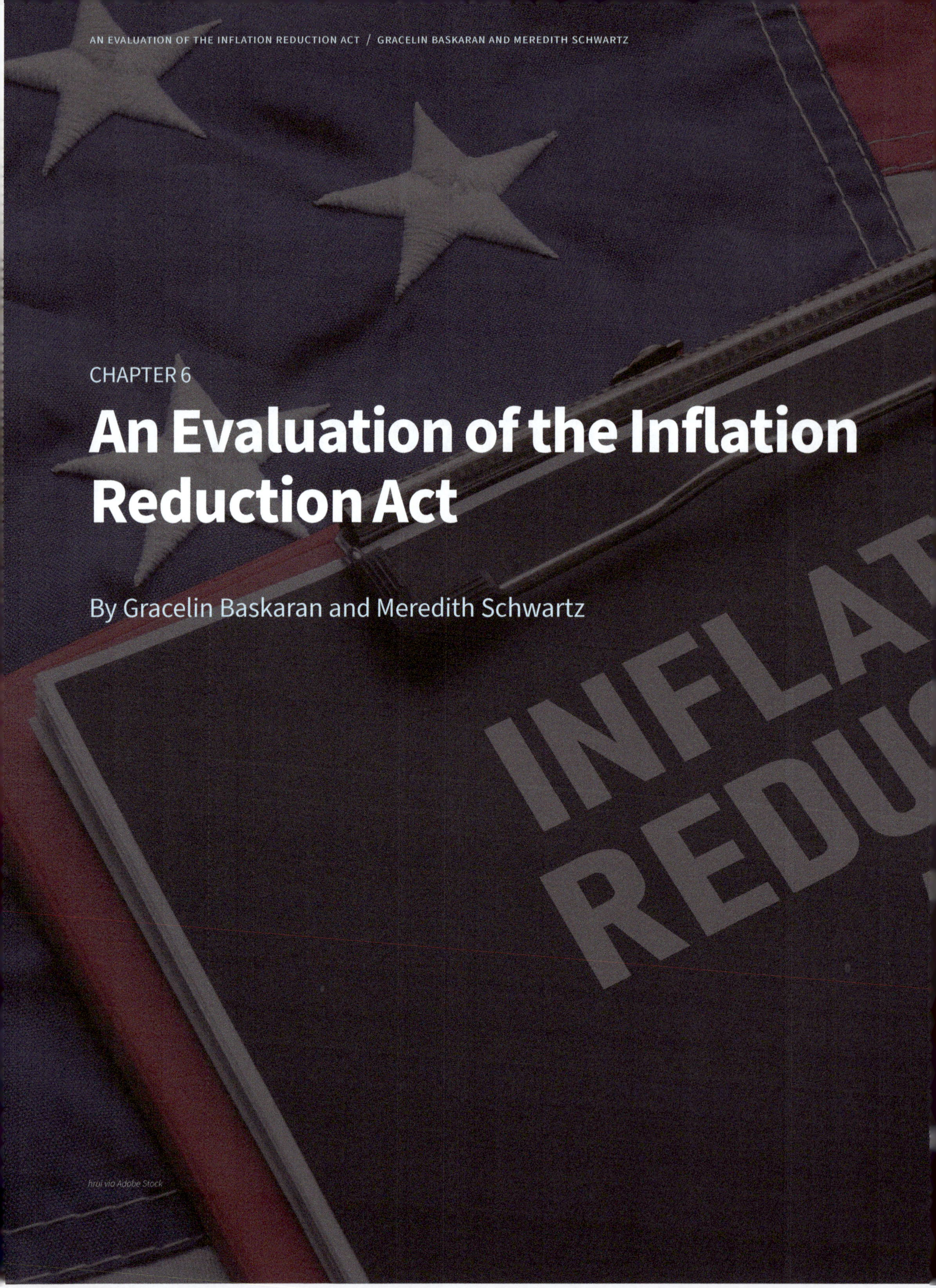

hrui via Adobe Stock

Supply chain shifts and industrial transitions take time, but with the right adjustments, IRA incentives have the potential to catalyze shifts in mineral supply chains that may change the entire security landscape

The Inflation Reduction Act (IRA) of 2022, the Biden administration's flagship piece of legislation, aims to address climate change by boosting U.S. manufacturing of clean energy technologies such as electric vehicles (EVs), solar panels, and wind turbines, as well as their components—including critical minerals. The Biden administration has set ambitious targets of cutting all carbon emissions from the transportation sector and achieving net-zero carbon emissions by 2050.[174] To help meet these goals, the IRA was enacted to rapidly accelerate the development and uptake of EVs, as well as renewable energy projects, to decarbonize the electrical grid. All of these technologies are highly minerals intensive.

The IRA was a politically contentious piece of legislation at the time of its passage, garnering exclusively Democratic support—not a single Republican in the House or Senate voted for the bill.[175] However, that has not stopped Republican-led states from benefitting from the incoming investment, jobs, and technical skills spurred by the legislation. As of August 2024, nearly 85 percent of the total $126 billion in investments and 68 percent of the 150,000 newly created jobs have flowed to Republican districts.[176] Ohio, North Carolina, and Georgia have been major recipients of IRA-induced projects, which may signal the beginning of a manufacturing renaissance accompanying the clean energy transition.

The bill's provisions of tax credits, grants, and loan guarantees have led to a boom in domestic clean energy projects, catalyzing investment from abroad into U.S. manufacturing of EV batteries and solar cells. As these projects take off, the demand for critical minerals such as lithium, cobalt, and nickel is projected to rise.[177] Securing supply chains for minerals like lithium, nickel, cobalt, graphite, and rare earth elements (REEs) is not only a goal of the IRA but also vital to the success of the law's broader decarbonization objectives. The Biden administration has repeatedly emphasized the importance of securing critical minerals through the IRA. However, the implementation of the IRA's provisions has been insufficient to fully address the needs of critical minerals supply chains.

To understand the IRA's impact on critical minerals security, this chapter describes each relevant provision and assesses its strengths and weaknesses. The IRA is then evaluated on its success in achieving its intended goals: accelerating decarbonization, boosting domestic manufacturing, and ensuring the security of critical minerals supply chains independent of adversaries. Analyzing the IRA provides valuable insights that can be applied to the mineral supply chains of other industries, including energy technologies, semiconductors, defense applications, and the industries that drive the modern economy. Landmark legislation like the IRA has the potential to make or break U.S. critical minerals security goals.

WHAT PORTIONS OF THE INFLATION REDUCTION ACT ADDRESS CRITICAL MINERALS?

The IRA creates or expands the following incentives that apply to the critical minerals industry: Section 30D New Clean Vehicle Credit, Section 45X Advanced Manufacturing Production Credit, Section 48C Qualifying Advanced Energy Project Credit, additional capital for the Defense Production Act Title III, and the programs of the Department of Energy's Loan Programs Office.

Section 30D New Clean Vehicle Credit

Section 30D of the IRA incentivizes the procurement of minerals for battery manufacturing from allied sources by providing up to $7,500 in tax credits for

Figure 1: Free Trade Agreement Countries

Source: CSIS analysis

qualifying EV purchases. This provision is key to increasing affordability and deploying EVs at scale. To be eligible, vehicle manufacturers must meet several criteria, including sourcing requirements for critical minerals and components. The Department of the Treasury ruled that EVs will only be eligible for $3,750 of the credit if they meet specific thresholds for the critical mineral content of their batteries. As of 2024, 50 percent of the minerals in an eligible vehicle must be extracted and processed in the United States, a country with which the United States has a free trade agreement (FTA), or Japan, which is the sole beneficiary of a critical minerals agreement (CMA). This threshold increases annually, reaching 80 percent by 2027.

To receive the tax credit, vehicles must also comply with the foreign entity of concern (FEOC) rules. The Bipartisan Infrastructure Law defines an FEOC as any entity "owned by, controlled by, or subject to the jurisdiction or direction of a government of a foreign country that is a covered nation."[178] There are four FEOCs: China, Russia, Iran, and North Korea. The Department of Energy further clarified the definition in December 2023, determining that an entity is considered an FEOC if 25 percent or more of its voting rights, board seats, or equity interest are held by the government of a covered nation, or if the entity is effectively controlled by an FEOC through a license or contract.[179] To qualify for Section 30D benefits, vehicles must not contain any minerals mined or processed by an FEOC, as defined. If any supplier to a vehicle manufacturer is in violation, the vehicle will not qualify for the Section 30D tax credit. This FEOC requirement goes into effect on January 1, 2025.

These sourcing requirements were intended to catalyze investments in critical minerals production in both the United States and its free trade partners, as well as to incentivize auto manufacturers to shift away from foreign adversaries in favor of domestic mineral suppliers. However, an unintended consequence of the Section 30D tax credit has been the disqualification of many EVs manufactured in North America.[180] At the start of 2024, only 10 vehicles qualified for the credit—a number likely to decrease once the FEOC rules go into effect in January 2025.[181]

The FEOC threshold has come under rightful scrutiny. On the one hand, critics argue that the provision is too permissive and should not allow firms with any Chinese ownership to receive taxpayer-subsidized benefits from the IRA. On the other hand, the FTA threshold is viewed as too restrictive, as most FTA countries lack the minerals needed for EV battery manufacturing. There is broad consensus that legislative loopholes need to be closed, given that Chinese firms are reportedly using parent or shell companies to benefit from the tax incentives.[182]

Congressional leaders from both sides of the aisle have expressed concerns about the implementation of Section 30D, particularly regarding FEOC guidance. Senator Joe Manchin (I-WV), at the time a Democrat and one of the original sponsors and key architects of the IRA, led a bipartisan group of senators to introduce a Congressional Review Act resolution of disapproval aimed at overturning the Department of the Treasury's final rule implementing the Section 30D Clean Vehicle Credit. Manchin objected to the Biden administration's interpretation of the Section 30D sourcing requirements, describing it as inconsistent with the intent of Congress. The senator stated that the Department of the Treasury's determination "allows China to gain control of our nation's auto industry."[183]

Graphite is the greatest sourcing hurdle for most battery and EV manufacturers. It is the largest mineral component of a battery by weight, comprising over 145 pounds of the total 456 pounds of minerals in an EV battery. In comparison, an internal combustion engine requires no graphite and uses only 75 pounds of minerals in total.[184] China accounts for 77 percent of natural graphite production, over 95 percent of synthetic graphite production, and nearly 100 percent of graphite refining.[185] The United States, meanwhile, contains less than 1 percent of the world's graphite reserves and is 100 percent import reliant, as it has yet to develop a domestic graphite mine despite significant government support for Graphite One's project in Nome, Alaska.[186] To keep U.S. EV manufacturing on track, the Department of the Treasury exempted graphite from FEOC requirements for two years, declaring it "impracticable-to-trace" until 2027.[187]

The nascent North American graphite industry has raised concerns that the exemption could impede the development of a domestic graphite industry which cannot compete with Chinese prices.[188] While the graphite exemption is necessary for the U.S. auto industry to remain competitive in EV manufacturing, alternative sources of graphite outside China are developing rapidly and are expected to come online in the coming years. Therefore, the exemption should remain in place until 2027 but should not be extended further. This will incentivize automakers to source graphite from Western suppliers and eliminate reliance on FEOC graphite by 2027.

Given the disapproval surrounding Section 30D's implementation from both Congress and industry, the provision's longevity under a new administration remains politically precarious. For Section 30D to remain politically viable for future administrations, it must be implemented in a manner that better protects the domestic industry, includes more strategic international partners, and prevents Chinese firms from benefitting, as Congress intended.

Section 45X Advanced Manufacturing Production Credit

Section 45X of the IRA provides a 10 percent credit for the costs incurred during the production of critical minerals to specified levels of purity.[189] The credit was intended to expand domestic production of critical mineral components for clean energy technologies. Notably, the credit phases out for all industries by 2032, except for critical minerals projects. This means that, if properly executed, the Section 45X credit could provide perennial financial support to capital-intensive critical minerals projects.

However, in December 2023, the Department of the Treasury proposed a rule that "direct and indirect material costs . . . and any costs related to the extraction or acquisition of raw materials would not be taken into account as production costs."[190] This would mean the tax credit could only be applied to mineral processing, not mining. The credit was not optimally designed for mineral processors either. According to the proposed rule, the costs of attaining raw minerals would not be covered under the Section 45X credit, but these raw material costs are often the primary expenses for mineral processors and recyclers. Therefore, even eligible mineral processors were not gaining much financial support from the Section 45X credit.

The proposed Treasury Department rule met with intense opposition from private industry and congressional leaders. Nine Democratic senators wrote a letter to Treasury Secretary Janet Yellen in direct response to the department's Section 45X rulemaking, expressing their displeasure and contending that the rule does not align with congressional intent. The National Mining Association, representing members including General Motors, MP Materials, Perpetua Resources, Rio Tinto, South32, and a host of other mining and automotive companies, submitted public comments that, without the inclusion of material and extraction costs in the calculation of Section 45X credits, "the impact of the 45X credit is significantly reduced."[191]

In October 2024, the Department of the Treasury released its final rule on Section 45X.[192] In a stunning reversal, the final ruling recognized that previous guidance did not incentivize proportional investments in the minerals sector due to restrictions on what counted as production costs. The final rule clarified that material and extraction costs may now be included in calculating credits for projects producing eligible processed and refined critical minerals.[193] This significantly changes the scale of benefits that critical minerals projects can derive from the Section 45X credit. Projects that are vertically integrating supply chains by feeding extracted ore from Western mining operations to U.S. refineries can now apply the 10 percent credit to the vast majority of their production costs in perpetuity. This will help offset the immense costs mineral projects face in a market highly vulnerable to Chinese manipulation.

Section 48C Qualifying Advanced Energy Project Credit

The Section 48C investment tax credit provides $10 billion in credits for qualifying projects that fall into one of three categories:

1. **Clean Energy Manufacturing and Recycling:** Reequips, expands, or establishes an industrial or manufacturing facility for advanced energy properties (e.g., EVs, solar panels, or energy storage systems)
2. **Industrial Decarbonization:** Retrofits a manufacturing facility in an energy-intensive sector to reduce greenhouse gas emissions
3. **Critical Materials:** Reequips, expands, or establishes an industrial facility for the processing, refining, or recycling of critical materials

Manufacturers can seek a credit worth up to 30 percent of the total amount invested in facilities that produce critical minerals. Unlike the Section 45X credit, the Section 48C program is competitive—not all projects that apply for the credit will be granted benefits. In the first round of Section 48C awards, of the 250 full applications requesting $13.5 billion in credits, 35 projects totaling $4 billion in awards were selected. This is only a 14 percent acceptance rate.[194]

Of the projects selected for the credit, two-thirds fell under the clean energy manufacturing and recycling category, meaning that critical materials projects were deprioritized in the first tranche of funding.[195] So far, it seems the Section 48C tax credit has been meaningful in incentivizing companies to make investments in domestic production of critical materials, although funding remains more concentrated in downstream applications.

Applications for the second and final tranche of 48C funding were due in October 2024, and awards were announced in January 2025 to allocate the remaining $6 billion of funding.[196] The distribution of funds for the second tranche was similar to the first, with only 25 percent of awards going to critical materials. Unless further funding is appropriated by Congress, the Section 48C program will end, rendering future projects ineligible.

Minerals projects are at a disadvantage in meeting Section 48C timelines due to the condition that a project is only eligible if it has received all federal, state, and local permits within two years of receiving an award.[197] This timeline is challenging for minerals projects to meet, considering it can take decades to obtain all the necessary permits. By the time many minerals projects would be able to receive all needed permits, mobilize capital, and put together an application, Section 48C will have no remaining funds.

Defense Production Act Title III

The Defense Production Act (DPA) of 1950 was first enacted in response to the Korean War to expand the authorities of the president to ensure U.S. industry has the capability and capacity to meet national security needs. Title III of the act gives the president the authority to leverage incentives for private industry, such as loan guarantees, purchase commitments, and grants, to expand production capacity and supply of critical technologies.[198] Since 1950, the DPA has been routinely invoked by presidents to respond to a number of crises, including the Cold War, the energy crisis of the 1970s, and the Covid-19 pandemic. The DPA is set to expire in 2025 unless reauthorized by Congress.[199]

The IRA provided an additional $500 million toward the DPA Title III grants, which provide financial support to critical minerals projects that protect, expand, or restore industrial base capabilities critical to national security. Under the Biden administration, DPA Title III has been particularly important for building domestic midstream capabilities. To address this large vulnerability, DPA Title III has funded cobalt refining in Canada, titanium processing in North Carolina, and REE separation in Texas, along with several other critical minerals projects.[200]

Expansion of the Department of Energy's Loan Programs Office

Finally, the Department of Energy received an additional $11.7 billion in funding and was granted an additional $100 billion in loan authority for its Loan Programs Office (LPO). The office was first established in 2005 by the Energy Policy Act to fill the gaps in affordable private debt and bring promising energy technologies to market that would otherwise be unable to access private financing. In this way, the LPO acts as a "bridge to bankability" to kickstart projects that might otherwise take a long time to reach commercial scale. This is especially important for critical minerals projects that are struggling to attract financing during periods of depressed commodity prices.

The IRA expanded the scope of the LPO to create new programs and increased the loan authority by $100 billion. Critical minerals projects can qualify for funding under the LPO's Advanced Technology Vehicles Manufacturing Loan Program (ATVM) as a qualifying vehicle component.[201] If selected, companies can receive a direct loan or loan guarantee of up to 80 percent of the project's eligible costs.[202]

Some loans have been granted for mining and mineral processing projects. In 2022, Syrah Resources, an Australian company, received the first ATVM loan in over a decade and the first of its kind for the critical materials industry.[203] The $102 million loan from the Department of Energy will support its graphite processing facility in Vidalia, Louisiana. When Syrah's facility is complete, it will create the first vertically integrated graphite supply chain outside of China, sourcing graphite from its mining operations in Mozambique.[204]

Syrah Resources' graphite production, supported by the LPO, was once thought to be an exemplary use of IRA funds. With the government's support, the project was projected to boost U.S. manufacturing, create jobs, establish a new graphite supply chain independent of China, and create a graphite source that the U.S. EV industry can rely on to qualify for Section 30D benefits in 2027. However, in December 2024, Syrah Resources declared force majeure and suspended operations for its graphite mine in Mozambique, as unrest and violence plague the nation following a disputed election result.[205] Now, Syrah's Louisiana processing project is at risk as well, unable to source graphite feedstock from its partner project in Mozambique. The company's stock plummeted as Syrah defaulted on its U.S. government-backed loans.[206] This recent development highlights the inherent risk with LPO investments in projects with supply chains vulnerable to political disruption. The incident also shows the importance of diversifying mineral supply to decrease dependencies on a singular feedstock source.

Additional loans in the critical materials sector have been extended to Ioneer's Rhyolite Ridge project ($700 million) and Lithium Americas ($2.26 billion) for lithium production, as well as Li-Cycle ($375 million) for battery minerals recycling and recovery. But LPO grant funding for critical materials projects remains limited, with far greater amounts going toward battery manufacturing projects. Critical minerals projects seeking LPO funding still face many challenges—the LPO process is lengthy, challenging to navigate, and prone to delays, as upstream projects can take decades to obtain the permits and offtake agreements needed. While LPO funds would be meaningful for capital-intensive midstream projects, few currently exist in a market where achieving economic viability is so challenging.[207]

SCORING THE IRA'S CRITICAL MINERALS EFFICACY

The IRA has three primary objectives as they relate to critical minerals security: (1) accelerate the deployment of mineral-intensive clean energy technologies, (2) support domestic manufacturing, and (3) secure supply chains by eliminating dependence on foreign adversaries. The law's provisions described above were created with these objectives in mind. In practice, however, the IRA has yielded mixed results, and critical minerals supply chains have a long way to go before the United States and its allies have sufficient production to support the many clean energy projects underway downstream.

One of the IRA's largest shortcomings is in "friendshoring" critical minerals production. Section 30D was intended to incentivize U.S. EV and battery manufacturers to source critical minerals from the United States or its FTA partners. However, limiting the benefit to FTA partners excluded crucial allies and major minerals producers such as Argentina, Brazil, India, Indonesia, Namibia, Saudi Arabia, South Africa, Vietnam, Zambia, and the European Union. Meanwhile, the primary mineral production of FTA partners, such as Chile, Mexico, and Peru, is copper—a critical material for energy that is not even eligible for Section 30D benefits, given that it is not on the Department of the Interior's Critical Minerals List.[208] Therefore, Section 30D is largely ineffective for sourcing from partners abroad: most FTA partners are not substantially benefiting from the IRA due to their limited resource reserves, and mineral-rich countries that could benefit are excluded.

Table 1: Inflation Reduction Act Report Card

IRA Objective	IRA Impact	Grade
Accelerating Clean Energy Technology Development and Deployment in the United States		
Domestic critical minerals mining	Sections 45X and 48C ineligible, Section 30D, DPA III, and LPO eligible, but limited impact due to domestic permitting system	1.0
Domestic processing of mineral inputs	Processing projects spurred by Sections 45X and 48C, DPA, and LPO but magnitude of results remains to be seen amid challenging market conditions	3.0
Domestic manufacturing of final products (EVs, turbines, solar panels)	Many new factory announcements due to Sections 45X, 48C, and 30D, and LPO. However, EV sales remain depressed, and final products largely contain critical minerals from adversaries	2.0
Declining emissions[209]	Emissions fell 3 percent in 2023, largely attributed to increased renewable power in the electric power sector. Transportation sector emissions remained unchanged due to low EV uptake.	3.0
Category average		2.25
Boosting Domestic Manufacturing		
Job creation[210]	150,000 jobs created in 2023	4.0
Inbound investment[211]	$500 billion in total clean energy investment, many companies from Europe, Japan, and South Korea investing in large U.S. manufacturing projects	4.0
Category average		4.0
Securing Mineral Supply Chains Independent of Adversaries		
Export rerouting	Vertically integrated supply chains are still few and far between	2.0
Western offtake agreements	Section 30D is largely disincentivizing Western offtake of minerals in key non-FTA countries	2.0

Removing FEOCs from supply chains	Progress in REEs, graphite, and lithium production	3.0
Friendshoring mineral production	Some progress, but without CMAs or access to Section 30D benefits, investments to source minerals in non-FTA countries are disincentivized	2.0
Category average		2.25
Final Score		**2.83**

IRA Performance Key:

4.0 - Exceeding Expectations
3.0 - Meeting Expectations
2.0 - Below Expectations
1.0 – Poor

Source: CSIS analysis

One proposed workaround for granting more countries FTA status is through the use of CMAs, which are FTA-equivalent, minerals-specific agreements that could be negotiated on a shorter time frame than an entire FTA. The U.S.-Japan CMA was signed by the Biden administration in March 2023. However, the agreement was met with intense backlash from Congress, with Democrats as well as Republicans lamenting that the rushed agreement bypassed the role of Congress in ratifying FTAs.[212] Since then, no additional CMAs have been signed, and the list of FTA-equivalent countries sits at just 21.[213] As the critical minerals domestic sourcing requirement rises in the coming years, U.S. downstream manufacturers will need more diverse sourcing options to continue to qualify for benefits, but no framework to expeditiously incorporate more partners currently exists.

Without widening the beneficiary list, there is no incentive for Western producers to invest in projects in resource-rich countries, nor any incentive for these countries to seek Western offtake of their mineral production, as China remains a willing buyer. This is adversely impacting the United States. As the Pentagon invests in REE separating and refining capacity for permanent magnet technologies through its DPA Title III program, these projects will depend on a steady supply of ore feedstock. With less than 2 percent of the world's REE reserves inside U.S. borders, midstream refiners will need to work with global partners, like Brazil, to secure feedstock. Brazil holds nearly a fifth of the world's known REE deposits.[214] Brazil's first REE mine, Serra Verde, started commercial production in 2024, but without the qualifications for Section 30D benefits, Brazil has no incentive to sign a long-term supply contract with a U.S. firm.[215] Under the current IRA framework, the United States will lose offtake of key mineral deposits to Chinese refiners, and EV manufacturers will be unable to eliminate FEOC suppliers from their supply chains.

POLICY RECOMMENDATIONS

To make the IRA work better for mineral security, the U.S. government should pursue the following actions:

1. **The United States needs to build on the FTA partner list to include mineral-rich strategic partners currently excluded and left behind.** Section 30D's FTA requirements are arbitrary and were not created with mineral reserve locations in mind. The IRA cannot change geology, and global supply will remain reliant on Indonesia's nickel, Madagascar's graphite, and Argentina's lithium for the foreseeable future. To be less discriminatory, the U.S. government should, in consultation with Congress, come up with a list of approved countries for Section 30D credits that align with U.S. mineral security goals and human rights standards. This will incentivize a greater number of strategically important countries to work with Western partners over Chinese buyers.

   This task may seem daunting, but there are a number of bilateral and multilateral cooperative forums and trade agreements that could be expanded. For example, the Quadrilateral Security Dialogue, Indo-Pacific Economic Framework for Prosperity, and Americas Partnership for Economic Prosperity could become more useful if they included binding terms, such as investment incentives, eligibility for subsidies and concessional financing, and preferential market access. Trade agreements can also be revised. For example, the African Growth and Opportunity Act (AGOA) is up for reauthorization in 2025, and several of the beneficiaries—such as the Democratic Republic of the Congo, Madagascar, Namibia, South Africa, Tanzania, and Zambia—are rich in minerals. Most of these mineral exports currently go to China. By including an investment incentive—similar to Section 30D—in AGOA, the United States can both incentivize investment in these countries by Western companies and encourage routing supply to the United States instead of China.

2. **FEOC rules should be amended to phase out taxpayer-subsidized benefits for projects with any Chinese ownership.** At present, to qualify for IRA tax credits, mineral projects must be less than 25 percent owned by an FEOC. This means that a mine that is 24 percent owned by a Chinese firm in Peru would still receive IRA benefits, given that the United States has an FTA with Peru. As the United States and allied nations build their capabilities, the United States should steadily reduce the FEOC ownership threshold to zero percent. This does not mean that Chinese firms cannot export to the United States; it simply means they will not be eligible for the tax credits that give them a competitive edge.

3. **Additional subsidies, such as a price floor, are needed for mining projects both at home and abroad that feed U.S. processors and refineries.** Western mining and processing operations will not be able to compete against Chinese and Russian prices without additional government support. The cost of producing nickel at Chinese-owned Weda Bay in Indonesia is just $4.23 per pound, compared to the U.S. Eagle Mine's costs of $5.32 per pound.[216] China produces minerals using dirtier energy sources, cheaper labor, and government subsidies designed to price out Western competition. U.S. critical minerals producers need all the support they can get to compete as prices for key energy minerals reach two-year lows. A price floor would give mining companies and their investors assurance that their capital-intensive projects will not become economically unviable due to commodity price volatility.

   Another example of how price floors could be crucial for mineral producers is palladium production. Palladium is key to hydrogen fuel cells, but its production is primarily concentrated in Russia (43 percent of global production) and South Africa (36 percent). The United States produced only 3.5 percent of the world's palladium in 2023.[217] Russian palladium operations are substantially cheaper than those in South Africa and the United States due to low safety standards. The total cash cost for mining one ounce of palladium at Norilsk in Russia is just $402 per ounce, compared to $590 at South Africa's Mogalakwena operation.[218] Sibanye-Stillwater, a South African company, owns the only major palladium operation in the United States. Its operating costs for 2024 reached over $1,000 per ounce—greater than the selling price of palladium. At times, the company was losing as much as $600 per ounce produced.[219] As a result, the company announced nearly 800 layoffs for its palladium mining operations in Montana and paused production. A price floor could ensure that commodity prices do not fall below mining production costs, helping strategic mining operations remain operational and saving U.S. jobs.

CONCLUSION

The IRA is the most significant piece of legislation to date that addresses the component supply chain needed to secure clean energy technologies. It created a host of incentives designed to reroute mineral supply chains by stimulating investment in the United States and select partner nations. By some measures, the IRA has seen marked success; by others, there is still much work to be done.

Supply chain shifts and industrial transitions take time, but with the right adjustments, IRA incentives have the potential to catalyze shifts in mineral supply chains that may change the entire security landscape for the energy technologies of the future. To make the IRA work better for mineral supply chains, policymakers should expand the list of countries benefitting from mineral production and processing incentives; steadily tighten the FEOC ownership requirements to reduce the competitive advantage that firms with fractional Chinese ownership gain through the IRA; and leverage new financing tools, like price floors, to accelerate the development of mining and processing capabilities in the United States and allied nations.

The IRA serves as an important template for how incentives could be structured to secure critical minerals supply chains for a variety of industries beyond just energy. Just as the IRA targets the lithium, graphite, and REEs needed for EVs and wind turbines, similar legislative initiatives could target gallium and germanium for semiconductors or antimony and tungsten for defense technologies. Securing mineral supply chains for the modern economy is one of the great challenges of our time. To succeed, policymakers will need the right set of incentives to push the private sector to innovate and expand to its full potential.

CHAPTER 7

An Evaluation of the CHIPS Act

By Kellee Wicker

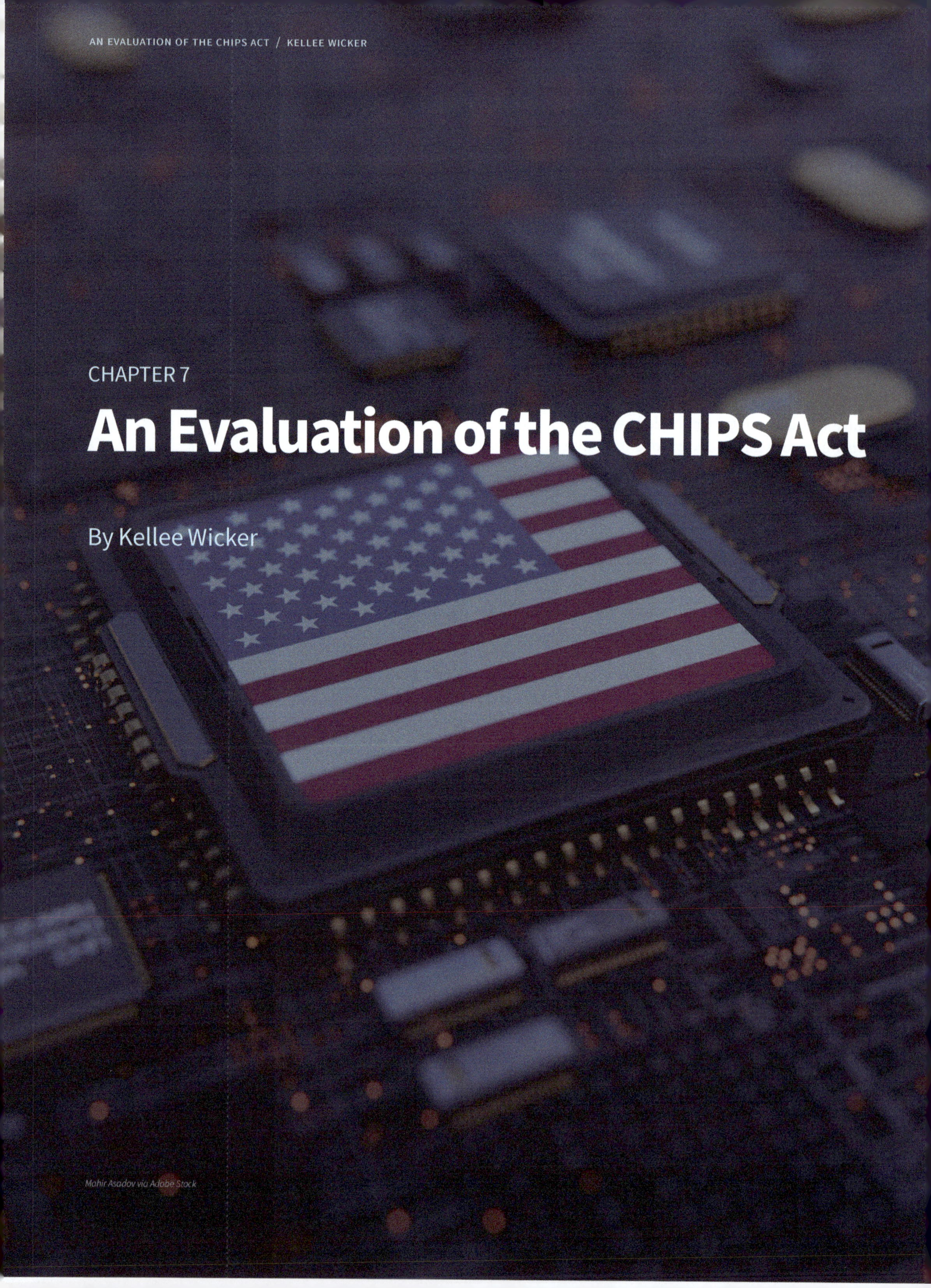

Mahir Asadov via Adobe Stock

As legislative and executive branch action on critical minerals has largely focused on minerals associated with clean energy and electric vehicles, Congress should specifically address the supply of semiconductor inputs.

Semiconductor design and production were once major sources of economic power for the United States. In 1994, U.S. semiconductor companies commanded 50 percent of the European market, 40 percent of the Asia-Pacific market, and 17 percent of the Japanese market. The electronics industry in 1995 was one of only seven manufacturing industries in the United States that employed more than one million workers, with 40 percent of those workers focused specifically on chips, and that percentage was rising.[220] At the time, Japan was the United States' primary rival for dominance in the global semiconductor industry, with the United States holding the largest share of the global market and Japan close behind, while South Korea and Taiwan were beginning to establish their footholds. Amid this competition, governments in the United States, Japan, and Europe invested strategically in public-private partnerships for research and development (R&D) and in maintaining domestic production, aiming to secure an edge in a highly competitive market.[221]

Today, the global semiconductor market is more valuable than ever, but the global supply chain has become concentrated. Furthermore, the U.S. share of fabrication has fallen to about 10 percent in 2020, although U.S. firms' market share remains robust due to their dominance in design and other key parts of the supply chain.[222] As noted in Chapter 2, this dependence on other nations for fabrication raised major alarms during the Covid-19 pandemic, when chip shortages rendered many consumer products unavailable or far more costly than usual. For possibly the first time, consumers became aware of the prevalence of chips in products from automobiles to refrigerators. U.S. policymakers also awakened to U.S. semiconductor dependence, particularly on Taiwan, which supplies over 60 percent of global chips and more than 90 percent of advanced chips.[223] Taiwan's position is precarious: while it gains some protection from its "silicon shield" as a major chips supplier, Taiwan remains a geopolitical thorn in China's side. Taiwan's connection to the world—and semiconductor consumers—could be severed at any time by China.[224]

The Creating Helpful Incentives to Produce Semiconductors Act of 2022 (also known as the CHIPS and Science Act, but hereafter referred to in this chapter as the CHIPS Act) is the United States' effort to invest in supply chain resilience. It aims to strengthen domestic production to guard against future chip shocks and—hopefully—limit China's ability to threaten U.S. access to semiconductors by addressing reliance on Taiwan and other suppliers, while also boosting investments in U.S. innovation and research. However, the act did not include any provisions addressing mineral access, leaving access to key minerals such as gallium and germanium highly vulnerable.

THE PATH TO THE CHIPS ACT'S PASSAGE

Following the pandemic supply chain shock, a number of semiconductor bills were introduced, many of which contained provisions that would eventually be passed in the CHIPS Act. The key bill was the CHIPS for America Act, introduced in the U.S. Senate on June 10, 2020, by Senators John Cornyn (R-TX) and Mark Warner (D-VA), along with a companion bill in the U.S. House of Representatives from Reps. Doris Matsui (D-CA) and Michael McCaul (R-TX).

But the CHIPS for America Act was not the only legislative vehicle proposed to counter China by strengthening domestic industries. Senators Charles Schumer (D-NY) and Todd Young's (R-IN) Endless Frontiers Act, introduced on May 21, 2020, aimed to strengthen U.S. competitiveness in emerging technologies through investment in research and regional technology hubs. The Endless Frontiers Act was subsequently replaced by the United States Innovation and Competition Act

(USICA), sponsored by Schumer and introduced on April 20, 2021. USICA quickly ballooned from a bill targeting research investment to a massive $250 billion bill intended to tackle China's rising power in emerging technologies across all of government. Cornyn and Warner's CHIPS for America components were rolled into USICA, along with several other competitiveness bills and programs authorized by the FY 2020 National Defense Authorization Act (NDAA). USICA passed the Senate in June 2021 with bipartisan support, but House counterpart bills, which included more progressive provisions, only passed along party lines. Differences between the versions passed by each chamber made reconciliation difficult, but eventually both chambers passed a House bill introduced by then Rep. Tim Ryan (D-OH) that had been amended to include significant portions of USICA along with key House demands on semiconductors, climate, and competitiveness.[225] The CHIPS Act of 2022 was signed into law on August 9, 2022, retaining a focus on both science and research capacity and securing semiconductor supply chain resilience.

Interestingly, despite the extensive list of provisions and goals included in earlier versions, critical minerals for semiconductors were not prioritized in the CHIPS Act, nor mentioned in media coverage or press releases as a priority for action. Securing U.S. access to critical minerals had been a key objective for the U.S. government, as seen in the Biden administration's February 2021 Executive Order on America's Supply Chains, which built on a September 2020 executive order from the Trump administration. This called for the secretary of defense to produce a report that would identify risks in the critical minerals supply chain and make specific policy recommendations.[226] Other bills passed in 2021, as outlined in other chapters of this volume, prioritized access to critical minerals for electric vehicles and renewable energy goals. The CHIPS Act did not address this interesting gap.

KEY CHIPS ACT FUNDS AND ACTIVITIES FOR SEMICONDUCTORS

The CHIPS Act appropriated $52.7 billion in funding for semiconductor-related activities through FY 2027, distributed across four distinct funds targeting different parts of the supply chain: the CHIPS for America Fund, the CHIPS for America Defense Fund, the CHIPS for America International Technology Security and Innovation Fund, and the CHIPS for America Workforce and Education Fund. These funds are appropriations for a number of programs and activities authorized by the FY 2021 NDAA, which had also been passed amid congressional concerns around U.S. semiconductor supply instability, as well as other activities.

The CHIPS for America Fund, through which the majority of the appropriated funds will flow, provides the Department of Commerce with $50 billion to incentivize domestic manufacturing capabilities, R&D, and workforce development. Of this total, $39 billion is earmarked for a program to promote investment in semiconductor fabrication, assembly, testing, advanced packaging, and R&D.

The remaining $11 billion of the CHIPS for America Fund is marked for R&D and workforce development programs. Chief among these is the National Semiconductor Technology Center (NSTC), which focuses on the research and prototyping of advanced semiconductor technology, including work on next-generation materials.

Another program, the National Advanced Packaging Manufacturing Program, is designed to enhance test, assembly, and packaging capabilities in the United States in coordination with the NSTC and the Manufacturing USA institutes. Lastly, funding is allocated for a National Institute of Standards and Technology (NIST) R&D program, which focuses on next-generation microelectronics, including materials characterization or the study of the structure and properties of materials, an essential component in understanding materials before exploring their potential uses.

The CHIPS for America Defense Fund appropriates $2 billion for the Department of Defense to stand up the National Network for Microelectronics Research and Development, also known as the Microelectronics Commons. The commons was created to support the "lab-to-fab" transition of chip innovation from academic development to commercial production and to expand the United States' edge on next-generation semiconductor development, including innovation in artificial intelligence, quantum technology, and telecommunications.[227]

The CHIPS for America International Technology Security and Innovation Fund (ITSI) appropriates $100 million annually for five years to the Department of State for coordination with foreign governments on telecommunications, semiconductors, and other emerging technologies.[228] In a briefing following the CHIPS Act's passage, the State Department unveiled its strategy for the ITSI fund, identifying "reliable access to critical minerals such as cobalt, aluminum, arsenic, copper, and rare earth elements" as one of its four priorities. The State Department would work to "secur[e] and diversif[y]" sources of these minerals across the entire processing chain, from mining to recycling.[229] However, with limited funds and three other priorities to balance across vast industries, the fund's impact on critical minerals projects is unlikely to be significant.

The CHIPS for America Workforce and Education Fund appropriates a total of $200 million for workforce development activities administered by the National Science Foundation (NSF). One part of that fund, National Science Foundation for the Future, does not explicitly focus on semiconductors but does touch on critical mineral dependence. The NSF director is tasked with supporting basic research "to advance critical minerals mining strategies and technologies for the purpose of making better use of domestic resources and eliminating national reliance on minerals and mineral materials that are subject to supply disruptions." Documentation on this fund also highlights the role of the NSTC's Critical Minerals Subcommittee in strategizing and coordinating federal efforts on critical minerals.[230]

THE CHIPS ACT'S IMPACT ON U.S. CRITICAL MINERALS SUPPLY CHAINS

Although minerals are referenced in various sections of the law, it is always as part of a list of desired outcomes and activities. As such, the CHIPS Act has had limited impact on critical minerals supply and access. However, future funding requests from fund beneficiaries could include projects that will either change domestic capacity for mining, processing, or recycling critical minerals or alter material needs for semiconductors.

The most identifiable direct impact comes as a result of ITSI funding for the Minerals Security Partnership, which convenes traditional allies such as the European Union as well as mineral-rich nations in Latin America and Africa.[231] This group has committed to environmental and social standards for mining projects in countries willing to work with the partnership, seeking to create less exploitative arrangements than those offered by China.[232] Ideally, this will lead to future resource agreements, reducing reliance on Chinese minerals, though timelines are long. As of March 2024, there are 23 confirmed projects across various critical minerals and rare earth elements, including gallium and germanium, but only two have reached key implementation milestones.

Another program with a clear intention to address critical minerals supply is the CHIPS for America incentives for semiconductor materials and manufacturing equipment facilities. The "Vision for Success" document for this program explicitly identifies reliance on China for upstream inputs like gallium as a supply chain resilience issue it seeks to address.[233] However, minerals are not the only inputs targeted, and the program's focus on companies looking to establish or expand their U.S. footprint may make it difficult for minerals companies to compete for funds. It remains to be seen whether minerals companies could—or would—successfully compete for this opportunity.[234]

Other impacts on critical minerals seem largely theoretical to date, although tracking funding disbursed through numerous organizations and mechanisms is challenging. The CHIPS for America Vision for Success document on incentives for commercial fabrication facilities stipulates that CHIPS funding applicants must also attract associated suppliers, including "reliable" material suppliers "committed to operating and innovating in the United States."[235] The document also highlights fabrication companies' use of new materials with unique performance properties as a focus of the theme of reducing costs.[236] This focus has facilitated the expansion and modernization of facilities like GlobalFoundries' 200 mm fab in Vermont, which enables the production of next-generation gallium nitride (GaN) chips—a win for domestic production of cutting-edge semiconductors.[237] However, these chips still rely on critical minerals, and it is unclear whether CHIPS Act funding has spurred companies like GlobalFoundries to identify or invest in new partnerships to source minerals such as gallium.

Similarly, it is difficult to tell what impact funding for the Microelectronics Commons network could have on minerals. The network recently issued its first round of project awards but has released few details on what these "lab-to-fab" projects are.[238] Information from the Midwest Microelectronics Consortium, one of the regional hubs receiving project awards, suggests the projects are focused on prototypes that advance packaging, processing, or other functionalities rather than new materials, architectures, or techniques that impact mineral reliance.[239] However, the Northeast Microelectronics Coalition, another network hub, announced having won an award for the transfer of "High AlAl GaN from Lab to Fab," indicating a potential new development in aluminum gallium nitride use.[240] Otherwise, the network seems focused on other areas of work, such as AI, quantum technology, and electromagnetic warfare.

RECOMMENDATIONS FOR A STRONGER FUTURE

As legislative and executive branch action on critical minerals has largely focused on minerals associated with clean energy and electric vehicles, Congress should specifically address the supply of semiconductor inputs. Finding or creating domestic supply is no easy task, and any impactful investment in mineral access will require support for R&D to discover new refining methods, new sources, and other creative solutions. China had decades to subsidize and incentivize its domestic metals industry to dominate gallium production; the United States has much less time and faces greater limitations in terms of what its people are willing to accept in terms of economic and environmental costs.[241] To address this challenge, the United States needs to focus on two fronts: developing new methods for mineral production, refinement, and recycling, and developing new approaches for semiconductor fabrication. Both, however, will take time.

These exigencies may serve as a necessary push for the semiconductor industry to step into a new phase of fabrication. Chipmakers are already concerned about reaching the maximum transistor density allowable by current two-dimensional production methods. Potential three-dimensional methods, such as carbon nanotubes, are seen by some as less viable due to higher costs and manufacturing difficulties compared to easier 2-D architectures.[242] While carbon nanotubes also depend on another critical mineral, graphite, they have already shown promise in various applications and could see wider deployment if cheaper alternatives become less viable—or there could be another production method yet to gain attention and investment.[243] Whether through a known but novel production method or a yet-to-be-industrialized one, the cost and time burden of adopting next-generation chip architectures, materials, or processes will only make business sense if current methods or materials become more expensive or unavailable.

Recycling

For Congress to effectively invest in identifying or expanding sources of gallium and germanium, it should prioritize incentivizing recycling and investing in research to improve recycling processes in the United States. The Department of Commerce's incentives program should be expanded to encourage private sector investment in building recycling facilities, using recycled materials, and exploring new methods of recycling gallium and germanium.

Current research on gallium and germanium processing from solid waste shows promise, with many researchers searching for more efficient recycling methods. One analysis of the field in early 2024 revealed several underused sources: gallium can be sourced from waste LEDs and dust from LED production, while germanium is recoverable from the significant waste generated by continuous technological upgrades, as well as wastewater from processes like fiber-optic cable production. Researchers are testing and evaluating various recycling and extraction methods, some with lower environmental impact, others with improved efficiency, and some that enhance the ability to extract multiple types of critical minerals from polymetallic resources (resources containing more than one critical mineral).[244] Notably, much of this research appears to be conducted by Chinese scientists and research teams, funded by government and higher education grants. It is vital that the United States directs research funding toward this work, with an emphasis on environmental standards and worker safety.

Cleaner, Safer Extraction

Additionally, funds should be appropriated for the NSTC and NIST to explicitly support research into cleaner mining and processing methods for gallium and germanium. For the Minerals Security Partnership or other initiatives to scale up nonexploitative and environmentally friendly mining to succeed, the United States must lead in finding economically practical methods that fulfill these objectives. While incentivizing private companies to conduct this research is possible, enabling higher-education institutions and other tech research hubs to make these discoveries increases the likelihood of broader benefits beyond a single corporation. Currently, private companies typically use proprietary practices for mineral processing, making it difficult to understand the factors controlling recovery and how processes could be improved.[245] If funding is appropriated for a private sector intervention, building in incentives for companies to not only enhance these practices but also share knowledge could be impactful. As with the current CHIPS Act funding, receiving these funds should be contingent on applicants' plans to base their operations in the United States or a friendly country and their commitment to positive environmental and labor practices.

Building Chips with Next-Generation Materials

Identifying the materials or combinations that will facilitate the next breakthrough in chip technology is a long-term project. Several types of investments will support the research needed to achieve this, but these discoveries can often come from unexpected fields. Beyond specific investments, the United States should broadly invest in basic R&D, focusing on areas the market may not fund but that, over time, could lead to paradigm-shifting discoveries.

Within the CHIPS Act's "family" of agencies and projects, additional funding explicitly aimed at identifying new materials or architectures using more common materials would be impactful. CHIPS for America's Metrology Program is already funding research on measurements—a key part of materials characterization—but it received limited funding under the CHIPS Act. Introducing an explicit research area for materials or architectures to the Microelectronics Commons would automatically encourage higher education and commercial institutions across multiple states to focus on this field while also emphasizing the commercialization of any new discoveries. The Microelectronics Commons offers the added benefit of allowing institutions to contribute their expert insights, replacing incomplete government guidance with field-specific expertise. Additionally, allocating specific funding for advanced materials research to NIST's CHIPS Act funding would provide fresh resources for the U.S. government's ongoing materials science efforts.

In the future, CHIPS Act–style funding should incentivize companies to adopt new processes and architectures that utilize novel materials. While market pressures likely will drive the adoption of discoveries that improve the efficiency of critical minerals use, advancing the next generation of semiconductor manufacturing will require government investment. This includes funding research to discover new materials, processes, or architectures and supporting the lab-to-fab transition for these technologies.

CONCLUSION

To date, U.S. legislative action on critical minerals has been tentative in addressing minerals relevant to semiconductor production. The process of identifying alternative sources for gallium and germanium will be expensive, likely slow, and possibly painful, but it is necessary to reduce the United States' 100 percent dependence on imports, particularly from China. The United States must prioritize recycling of technological waste and mining byproducts, invest in research for more efficient extraction and recycling processes, and work with companies already active in these industries. Basic R&D investment will also be key to transitioning from this period of uncertain and costly mineral access to a leap forward into the next generation of semiconductor fabrication and architecture—if such advancements are indeed possible.

CHAPTER 8

An Evaluation of the Defense Production Act

By Chris Michienzi

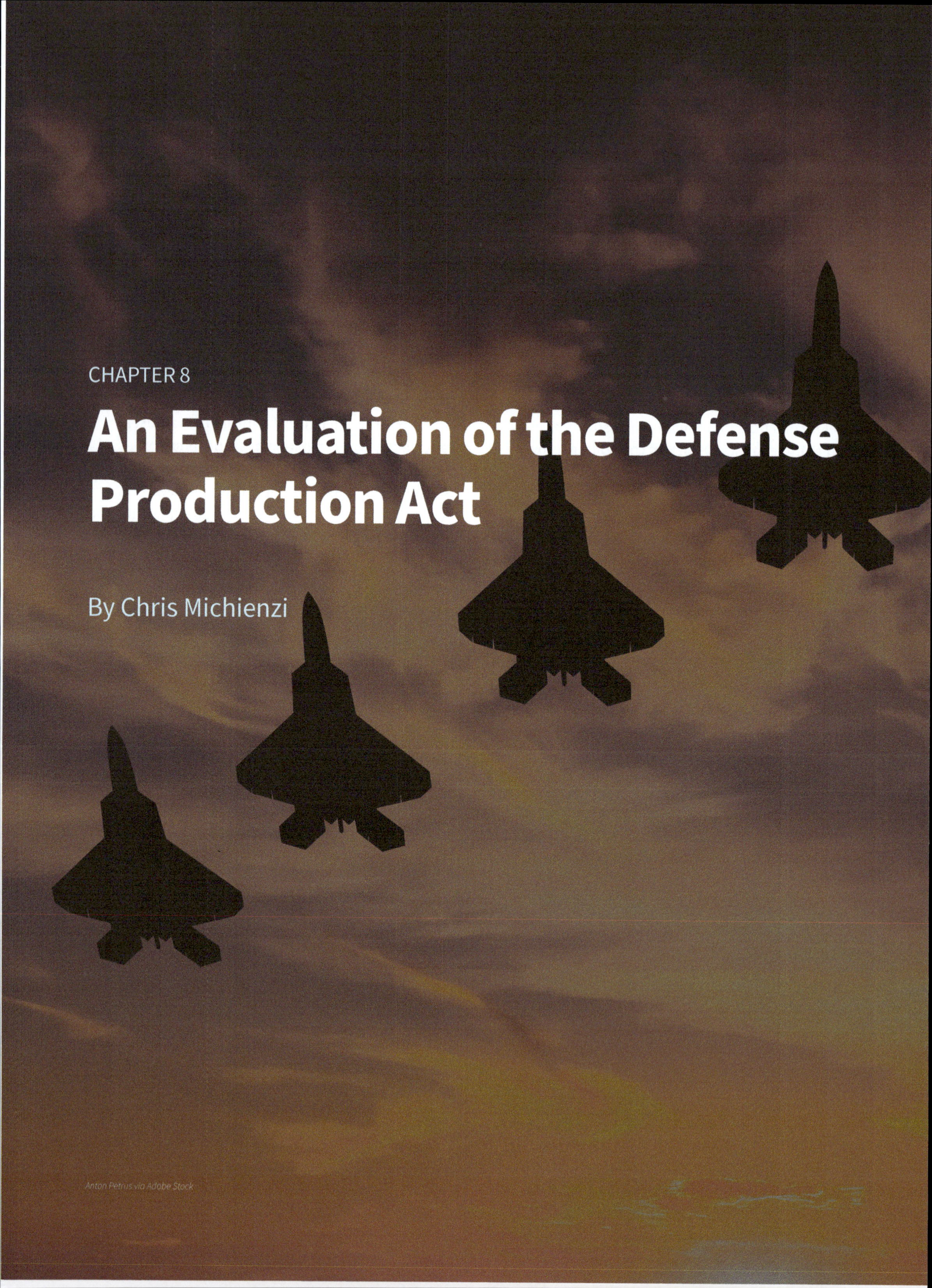

Anton Petrus via Adobe Stock

Critical minerals are used ubiquitously in commercial and defense systems and are therefore vitally important to national and economic security.

Critical minerals are used ubiquitously in commercial and defense systems and are therefore vitally important to national and economic security. On February 24, 2021, President Joe Biden issued Executive Order (EO) 14017, which called for a comprehensive review of critical supply chains in key sectors—one being critical minerals and other identified strategic materials, including rare earth elements (REEs).[246] The first deliverable of the EO required four government agencies to submit 100-day reports on four different sectors. The Department of Defense (DOD) wrote and published the report on critical minerals and strategic materials, with input from multiple other agencies, on June 8, 2021.[247]

The second deliverable of the EO called for the DOD and other agencies to write a one-year report on their supply chains. The DOD's report, Securing Defense-Critical Supply Chains, was published in February 2022.[248] It identified five key focus areas that were at high risk for the DOD, mostly due to adversarial foreign dependence within those supply chains. Critical minerals and strategic materials comprised one of the five areas, and the one-year report provided an update on the implementation of the recommendations from the 100-day report.

The DOD's Industrial Base Policy (IBP) office authored both reports. That office also manages two investment programs designed to mitigate shortfalls in the defense industrial base: the Defense Production Act (DPA) Title III program and the Industrial Base Analysis and Sustainment (IBAS) program. Since the publication of the EO 14017 one-year report, IBP has used the report's recommendations as a framework for making investments with both programs. Many of those investments are for onshoring capabilities that the United States has lost or for leveraging the capabilities of allies and partners.

The DOD's IBP office also authored the report in response to EO 13806, Assessing and Strengthening the Manufacturing and Defense Industrial Base and Supply Chain Resiliency of the United States, published in September 2018, and the Annual Industrial Capabilities Report to Congress, submitted yearly.[249] Both reports also detailed the fragility of the critical minerals supply chain; however, it was not until the EO 14017 report that the IBP office purposely aligned its investment programs for critical minerals with that report's recommendations. Additionally, the DPA Title III base budget significantly increased starting in FY 2021, allowing for increased investments to be made.

The Defense Production Act of 1950 authorizes the president to ensure the availability of U.S. and Canadian industry for U.S. defense, essential civilian, and homeland security requirements. The DPA is required to be reauthorized every few years, allowing Congress to make changes, such as expanding it to include space activity in 1975 and designating energy as an essential material good in 1980. Emergency preparedness during natural disasters or emergencies was added in 1994, and in 2003, the definition of national defense was expanded to include "critical infrastructure protection and restoration."[250]

Currently, the DPA only has three active authorities (although it has had more in the past): Title I, Title III, and Title VII.

Title I is for defense priorities and allocations, and includes provisions for the DOD to

- prioritize federal contracts over all other orders;
- control distribution of scarce materials within the civilian economy;
- allocate scarce materials against federal or private contracts; and
- prevent hoarding of scarce materials.

Figure 1: Defense Production Act Title III Budget and Funding

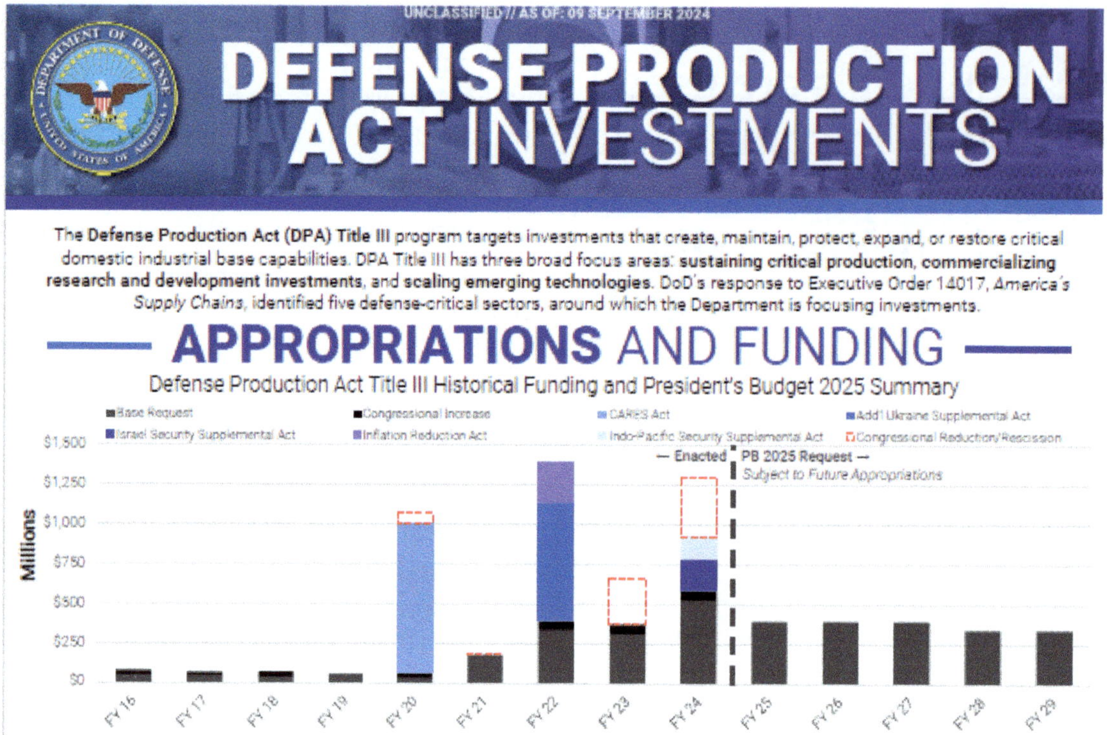

Focus Area	FY22 ($M)	FY23 ($M)	FY24 ($M)	FY25 ($M)
Kinetic Capabilities	55	172	206	84
Microelectronics	185	99	39	33
Strategic and Critical Materials	85	24	-	-
Castings and Forgings	5	38	41	-
Energy Storage and Batteries	-	16	-	-
Other Efforts	58	24	302	276
Total	**388**	**373**	**588**	**393**

Source: "Defense Production Act Investments," U.S. Department of Defense, September 30, 2024, https://www.businessdefense.gov/ibr/mceip/docs/DPAI-Infographic_30SEP2024.pdf.

Title III is for expansion of productive capacity and supply, which includes incentives for the DOD to develop, maintain, modernize, and expand production capacity for critical technologies via

- loans and loan guarantees;
- purchases and purchase commitments; and
- grants and subsidies.

Title VII is titled "General Provisions" and includes authorities for the DOD to facilitate and participate in

- antitrust immunity for industry, to develop and implement national emergency preparedness plans; and
- Committee on Foreign Investment in the United States (CFIUS) assessments.

DEFENSE PRODUCTION ACT TITLE III

This section focuses exclusively on DPA Title III. Grants and subsidies are the most common use of this title. The three priority areas for DPA Title III investment are (1) sustaining critical production, (2) commercializing research and development efforts, and (3) scaling emerging technologies.

DPA Title III has made investments in the United States and Canada, which is considered a "domestic source" for the purposes of the DPA. Canada has significant mining and material processing capability and exports a variety of strategic and critical materials to the United States.[251] The FY 2024 National Defense Authorization Act gave DPA Title III the ability to consider Australia and the United Kingdom as domestic sources as well.[252] This consideration could be critically important, as Australia is rich in mineral resources and has vast deposits of a large variety of critical minerals.[253]

Any investment made by DPA Title III requires the president, on a nondelegable basis, to identify a domestic industrial base shortfall as meeting three specific criteria:

- The industrial resource, material, or critical technology item is essential to national defense.
- Without presidential action under 50 U.S.C. §4533, U.S. industry cannot reasonably be expected to provide the capability for the needed industrial resource, material, or critical technology item in a timely manner.
- Purchases, purchase commitments, or other action pursuant to 50 U.S.C. §4533 are the most cost-effective, expedient, and practical alternative methods for meeting the need.

If these criteria are met, the DOD assembles a package with justifying information and sends it to the White House for approval. Once signed by the president, it is known as a presidential determination (PD). PDs are non-expiring, can be leveraged for different projects addressing the same shortfalls, and vary in breadth and scope depending upon the shortfall or challenge addressed. PDs are not an appropriation or funding mechanism, nor a mandate to address a specific shortfall or pursue a specific course of action. Examples of existing PDs related to critical minerals include the "Defense Production Act Title III Presidential Determination for Critical Materials in Large-Capacity Batteries," signed by President Biden on March 31, 2022, and five PDs for REEs, signed on July 22, 2019—two for separation and processing capability (light and heavy), one for metal and alloy processing capability, and two for rare earth permanent magnets production (samarium cobalt and neodymium iron boron).[254]

Interestingly, the same criteria necessary to obtain a PD explain why DPA Title III is an important mechanism for investing in critical minerals supply chains. DPA Title III funds are not to be used if other funding, for example, private investment or funding from other agencies, can be secured. The private sector is reluctant to make investments in critical minerals projects due to market price volatility for these materials. With China controlling most of the market and market pricing, many domestic and allied sources have been driven out of business—not exactly an attractive business case for private investors.[255] While the Department of Energy (DOE) has been making significant investments in this sector, it has been focusing on areas that align with its mission, as outlined in their vision statement: supporting "clean energy transition and decarbonization of the energy, manufacturing, and transportation economies, while promoting safe, sustainable, economic, and environmentally just solutions."[256] This leads to investments in projects for alternate designs that require no critical materials, more efficient and environmentally friendly mining, and reclaiming minerals through recycling. While the DOD is in favor of and has also been investing in these types of projects, many of the critical minerals necessary for national security applications must still be mined and processed using traditional methods, so the DOD is making those investments using DPA Title III.

Typically, funding is appropriated into the DPA Title III account annually, without specification on how it is to be spent, although Congress must be notified of planned expenditures. However, in the past few years, there has been an increased interest in the DPA from the executive and legislative branches and other government agencies, who are increasingly viewing DPA authorities as valuable tools to be leveraged against urgent, critical issues. For instance, in FY 2022, the DPA was appropriated $600 million by the Additional Ukraine Supplemental Appropriations Act, part of which included expanded domestic capacity for strategic and critical minerals.[257] Additionally, in FY 2022, the Inflation Reduction Act appropriated $500 million for enhanced use of the DPA. The funds were split equally between the DOE and DOD, with $250 million provided to the DOD to be applied to expanding capabilities for domestic mining, mineral processing, and related industrial sectors for large-capacity batteries.[258]

DPA Title III typically has a requirement for cost sharing from the companies receiving an award. The expectation is 50 percent of the total investment, but can be less depending on various factors, such as a company's ability to fund the total cost-share amount. Some smaller companies, such as small businesses and startups, do not have the cash flow available to fund a 50 percent cost share, and in those cases, DPA Title III may absorb a larger percentage of the total cost of the project. However, such instances may result in less DPA Title III funding available to make other investments.

The program also seeks to ensure that any investment it makes will lead to a sustainable capability not requiring further DPA Title III funding. One way to do this is with the seldom-used authority of purchases and purchase commitments. This allows the DOD to purchase or commit to purchasing a certain amount of a company's output over a set time period to ensure it can sustain the capability until market demand is sufficient to do so. This guaranteed demand also gives industry the confidence it needs to make its own investments. Purchases and purchase commitments were more easily executed previously because DPA Title III funds were appropriated as "no-year money," meaning the funds never expired; however, during the past three years, Congress has appropriated those funds as standard procurement, which expire in two years, making these longer-term commitments nearly impossible. This change was due to some members' concerns over timely execution rates; however, these concerns were unfounded, as the DPA Title III program executed $850 million of its $968 million budget in FY 2024.[259]

In order to offset some of the additional costs to the DPA Title III program and to help increase the demand for the newly developed capabilities, the program has recently developed its Pathfinders program, which "takes a revolutionary approach, focused on harnessing private capital and market forces to serve as a massive force multiplier to government investment."[260] The Defense Business Accelerator (DBX) program rapidly scales emerging technologies from the defense industry as well as DOD labs by leveraging the commercial market to accelerate production at scale. According to the DOD, "Since it was announced last fall, companies that received DBX awards are reporting they have received an additional $46.6 million in private capital, largely due to the spark provided by the initial $9.6 million DBX investment."[261] The Defense Market Catalyst (DMC) program is a public-private partnership (PPP) that stimulates trusted private capital for defense-oriented small businesses seeking to rapidly scale their operations to meet defense market needs. This capital infusion may also help those companies meet their cost-share obligations. And the Scaling Capacity & Accelerating Local Enterprises (SCALE) project looks at emerging technology through the lens of supply chain gaps, generating market pull and building PPPs to grow resilient domestic businesses.

INDUSTRIAL BASE ANALYSIS AND SUSTAINMENT PROGRAM

The Industrial Base Analysis and Sustainment (IBAS) program (10 U.S. Code §4817 established the Industrial Base Fund) was established in 2011 and has the following authorities:

- to support the monitoring and assessment of the industrial base
- to address critical issues in the industrial base relating to urgent operational needs
- to support efforts to expand the industrial base
- to address supply chain vulnerabilities

The only changes to the authority have been administrative in nature, such as amending the language to reflect the change from the "undersecretary of defense for acquisition, technology, and logistics" to the "undersecretary of defense for acquisition and sustainment." Congress does not prescribe how the base funding for the IBAS program (roughly $11 million in FY 2024) is to be spent; however, in the past several years, Congress has added a large amount of funding (around $1 billion in FY 2024) for specific industrial investment areas, for example, $175,692 for critical minerals in FY 2024. The IBAS program has always had the ability to and has made investments in all of the National Technology Industrial Base countries—Australia, Canada, and the United Kingdom.

Both the DPA Title III and IBAS programs have been making multiple investments in the critical minerals sector and have been investing in upstream and downstream capabilities to build and secure domestic critical minerals supply chains.[262] Table 1 shows all the critical minerals investments that have been made during the Biden administration.

Investments have been made in a multitude of

Table 1: DPA Title III and IBAS Critical Minerals and Materials Investments (2021–24)

Project	Awarding Program	Company	Location	Awarded ($, millions)	Date Awarded
Establish Comprehensive Domestic Tin Processing for National Hardware	DPA	Nathan Trotter & Co., Inc.	PA	19.0	Sep. 2024
Develop and Expand Production of Terbium Oxide from Recycled Fluorescent Light Bulbs	DPA	Rare Earth Salts	NE	4.2	Sep. 2024
Critical Minerals Workforce Development Project	DPA	Montana Technological University	MT	11.8	Aug. 2024
Ontario Cobalt Sulfate Refinery Project	DPA	Electra Battery Materials Corporation	CAN	20.0	Aug. 2024
Accelerated Development of the Thacker Pass Project	DPA	Lithium Nevada	NV	11.8	Aug. 2024
Accelerated Access to Domestic Manganese Ore for Advanced Materials Assessment Project	DPA	South32 Hermosa	AZ	20.0	May 2024
Expanding Domestic Capacity and Production of Cobalt for the Battery Supply Chain	DPA	Fortune Minerals Limited	CAN	CAN	May 2024
Expansion of Domestic Production Capability and Capacity of Natural Flake Graphite, La Loutre	DPA	Lomiko Metals Inc.	CAN	8.4	May 2024
Expansion of Domestic Production Capability of Nickel and Cobalt	DPA	The Doe Run Company	MO	7.0	Mar 2024
Upcycle Waste & Scrap to Prime Units for Critical Materials	DPA	6K Additive, LLC54	PA	23.3	Dec. 2023
Ceylon Graphite Project Bankable Feasibility Study	DPA	South Star Battery Metals Corp	CAN	3.2	Nov. 2023
Titanium Processing Plant	DPA	IperionX Limited	NC	12.7	Oct. 2023
Domestic Rare Earth Permanent Magnet Manufacturing Capability	DPA	e-VAC Magnetics, LLC	SC	94.1	Sep. 2023
Domestic Mining and Production of Lithium	DPA	Albemarle Corporation	NC	90.0	Sep. 2023
Advance Nickel Exploration and Mineral Resource Definition of the Tamarack Intrusive Complex	DPA	Talon Nickel (USA)	MI	20.6	Sep. 2023
Light Rare Earth Separation and Processing	IBAS	Lynas USA, LLC	TX	138.0	Aug. 2023
Developing a Domestic Advanced Graphite Supply Chain Solution Through the Graphite Creek Resource	DPA	Graphite One (Alaska)	AK	37.5	Jul. 2023
Feasibility Studies to Expand Cobalt Extraction	DPA	Jervois Mining USA	ID	15.0	Jun. 2023
High Purity Aluminum Capacity Expansion	DPA	Arconic Corp	IA	45.5	Jun. 2023
Rare Earth Element Separation Technology Capabilities Prototype Project	IBAS	Innovation Metals Corp.	CAN	4.0	Jun. 2023
Rare Earth Extraction from Acid Mine Drainage	IBAS	West Virginia University	WV	3.0	Jun. 2023
Heavy Rare Earth Separation and Processing	IBAS	Lynas USA, LLC	TX	120.0	Jun. 2022
Heavy Rare Earth Separation and Processing	IBAS	MP Materials Corp	CA	35.0	Feb. 2022
Rare Earth Elements Separation and Processing	DPA	Lynas USA, LLC	TX	30.4	Jan. 2021

Source: Data from Aissa Tovar, deputy director, DPA Title III, Office of the Secretary of Defense, Industrial Base Policy.

different critical minerals, including both light and heavy REEs, aluminum, cobalt, nickel, lithium, graphite, titanium, manganese, and terbium oxide. Projects range from mining (cobalt, manganese ore, and lithium) to REE separation and magnet making (NdFeB). They also include production of titanium, of which the DOD gets 95 percent from non-domestic sources, including Russia, and production of aluminum, both of which are used in the production of aircraft, ships, and submarines. There is also an investment in Chinese export-restricted graphite. Additionally, there are several recycling projects, including extracting REEs from acid mine drainage and production of terbium oxide from recycled fluorescent light bulbs. There is even a workforce development project that was awarded in August 2024. While these projects span a wide breadth of the critical minerals landscape, there are still some gaps. For instance, there are no investments in samarium cobalt magnets or for gallium or germanium, for which China recently restricted exports. This may be due to a lack of funding, as DPA Title III must fund many different industrial base sector shortfalls.

Figures 2 and 3 show the road maps for combined DPA Title III and IBAS current and future investments for strategic and critical materials and batteries, both of which contain projects for critical minerals. There continue to be investments in REE projects. Investments are also being made in niobium and tungsten, materials that go into refractory metals and superalloys and are used for components such as turbine engine blades and penetrators. Additionally, there are now projects for germanium and gallium, as well as additional recycling and workforce projects. Mining and refining of the main elements used in battery production—nickel, cobalt, lithium, and manganese—continue to be the focus.

Figure 2: Current and Future DPA Title III and IBAS Strategic and Critical Materials Investments

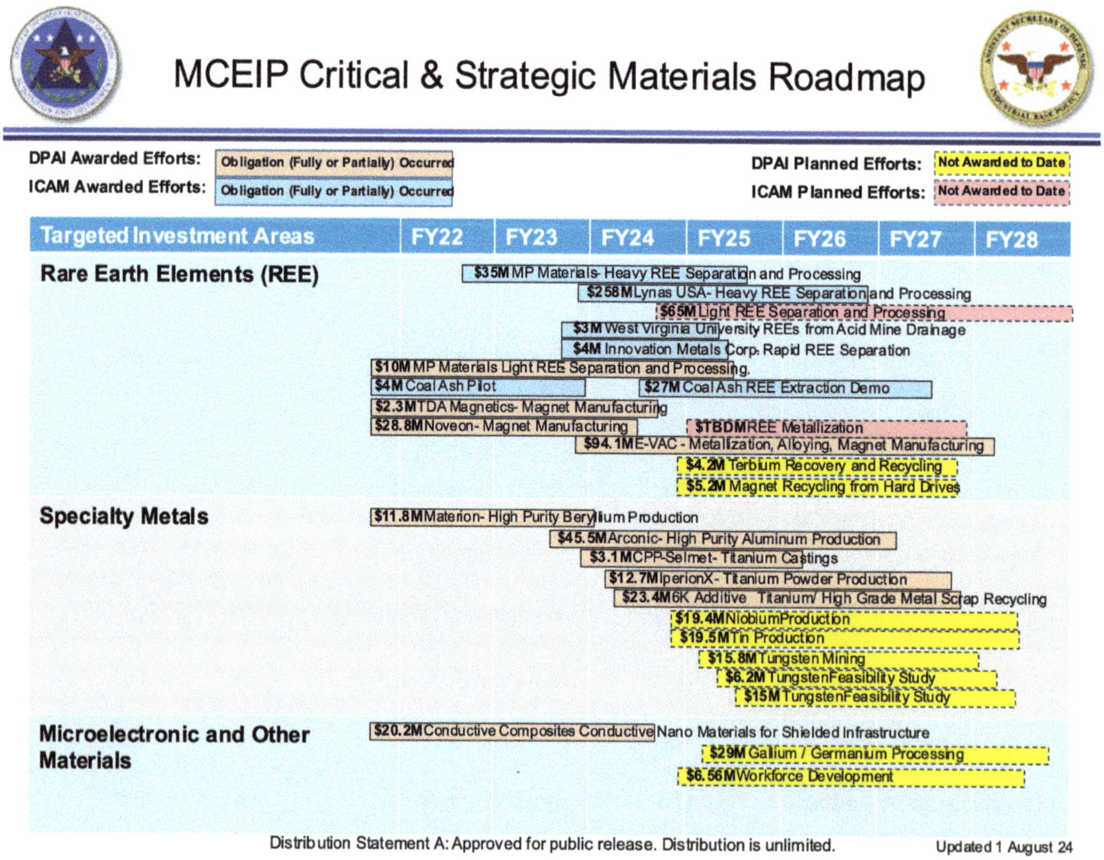

Source: "Manufacturing Capability Expansion and Investment Prioritization (MCEIP) Investment Roadmaps," Office of the Assistant Secretary of Defense for Industrial Base Policy, updated August 1, 2024, https://ndia.dtic.mil/wp-content/uploads/2024/eti/Tovar.pdf.

Figure 3: Current and Future DPA Title III and IBAS Energy Storage and Batteries Investments

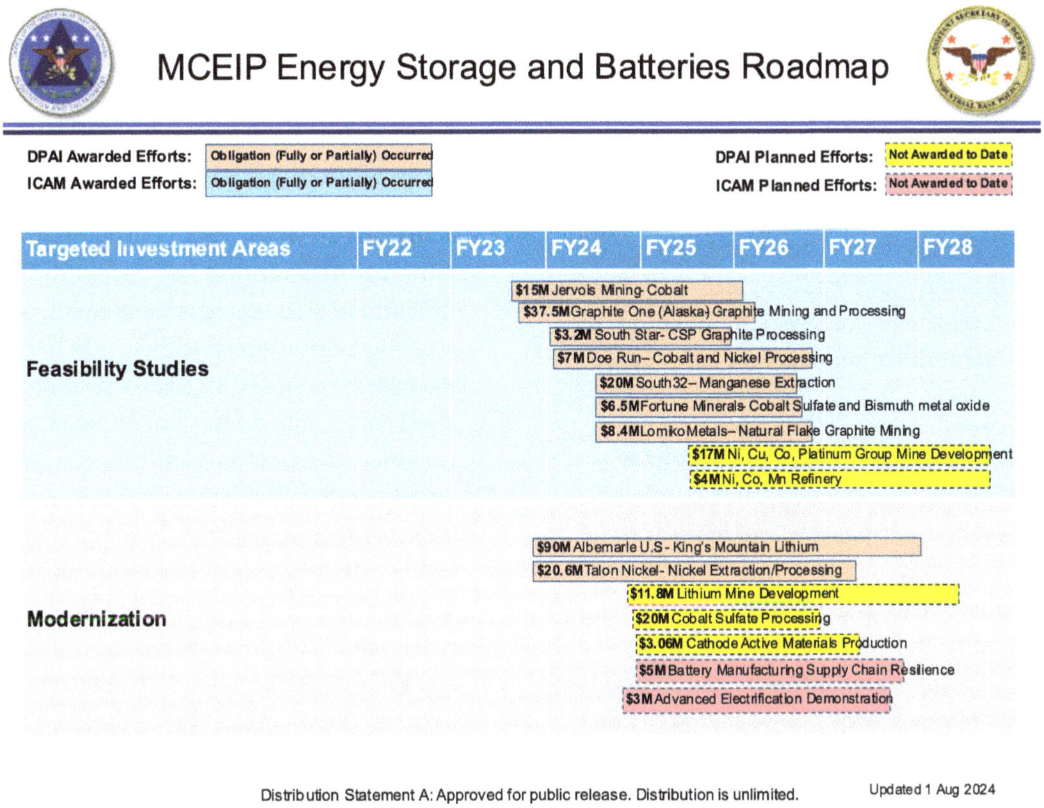

Source: "Manufacturing Capability Expansion and Investment Prioritization (MCEIP) Investment Roadmaps," Office of the Assistant Secretary of Defense for Industrial Base Policy, updated August 1, 2024, https://ndia.dtic.mil/wp-content/uploads/2024/eti/Tovar.pdf.

RECOMMENDATIONS

While the DOD has made significant investments—totaling $781 million—in the critical minerals sector to onshore or friend-shore capabilities and reduce U.S. dependencies on adversarial sourcing during the Biden administration, there is still much to be done. It has taken decades for these capabilities to atrophy, and it will take many years and many more millions of dollars to rebuild them. More investment dollars are needed, especially if the government's predicted timeline (by roughly 2027) for Chinese readiness for a potential invasion of Taiwan is to be believed.[263] If the United States were to come to Taiwan's aid, it is unlikely that China would continue to provide the country with critical minerals. Therefore, the capability for critical minerals production must be rebuilt quickly, before it is too late.

In addition, the DPA Title III budget is used to address a multitude of industrial base shortfalls, not just critical minerals. Even though the total budget may seem large, the amount available for critical minerals is not enough to address the considerable gap that exists.

Increased appropriations for both the DPA Title III and IBAS programs would allow for greater investments in critical minerals. The DPA Title III base budget increased from $373 million in FY 2023 to $588 million in FY 2024, but the program demonstrated it could execute much more ($850 million). Appropriating a base budget for DPA Title III that is at least $750 million to $1 billion would give the program the flexibility to invest more in critical minerals while maintaining or increasing investments in other important sectors. The IBAS program has consistently shown it can execute annual budgets at or near $1 billion, so appropriating a base budget in that range would help ensure consistency for critical minerals investments, rather than relying on inconsistent congressional additions.

There are several other recommendations that would help improve the ability of the DPA Title III program to impact the critical minerals landscape:

- **Delegate authority for approval of PDs to the secretary of defense.** Currently, it takes approximately one year to staff the PD package and obtain the president's signature. This timeline needs to be shortened to allow for more responsive industrial base investments. The secretary of defense is well positioned to understand the national security implications of the threat and can confer with the heads of other departments and agencies who may have equities, prior to signing a designation for the use of DPA Title III authorities for a particular critical minerals industrial base shortfall.
- **Make greater use of purchases and purchase commitments.** As mentioned earlier, purchases and purchase commitments will help sustain the capabilities that have received investments and provide industry with the demand signal and business case to make their own investments. Greater utilization of this authority will rely on two factors: an increased budget and a return to non-expiring funding being appropriated by Congress.
- **Increase investments in the workforce.** The workforce shortage is one of the key issues in all sectors of the industrial base, as was highlighted in the DOD's EO 14017 report. The IBAS program has been making investments in the workforce for several years, but the DPA Title III program has traditionally not made such investments, primarily due to a smaller overall budget. The IBAS program has been very successful with its National Imperative for Industrial Skills (NIIS) initiative, which facilitates multiple approaches to recruit, train, hire, and retain skilled workers. The NIIS initiative recognizes the interplay of K–12 and post-secondary education and training tracks. The key principles of the program are (1) active, sustained partnering with industry, academia, military departments, federal agencies, and state governments; and (2) creating regionally focused activities targeting capacity where skills are most needed (e.g., regional submarine industrial base efforts intensified and scaled in New England and Virginia). Additional funding would enable both programs to support additional workforce development efforts in key sectors, such as critical minerals.
- **Continue to leverage private equity and venture capital.** Private capital will multiply the impact of DPA Title III funding and assist small companies that lack the cash flow to fund cost share for awards. It will also mean DPA Title III does not have to absorb the extra cost, allowing those funds to be allocated to other projects.
- **Continue to work with commercial industry to generate market pull and demand.** The DOD's demand in this sector is very small, so it is vital that commercial industry helps generate part of the demand signal to develop resilient and sustainable domestic supply chains. For instance, the automotive industry also uses rare earth magnets in electric motors and sensors, as well as lithium batteries in electric vehicles.[264] Other critical commercial technologies, such as cell phones, are equally dependent on these materials.[265] While commercial industries may not view dependency on China from a national security perspective, they should be concerned from an economic security and liability perspective. In such cases, the DOD should continue to work with these industries to help provide the necessary demand signal to sustain a domestic capability for these materials.

The DOD recognizes the importance of critical minerals to national security. These minerals are present in nearly all the weapons systems and platforms the DOD purchases and uses. Although the department's demand is small compared to the commercial sector, the criticality of these materials drives it to make investments to ensure a secure supply. The investment strategy is derived from recommendations in the DOD's report in response to EO 14017, which identified strategic and critical materials as one of the five key focus areas at risk for the DOD.

The DPA Title III and the IBAS programs, both led by DOD's IBP office, have made and continue to make investments in the critical minerals industrial base. The IBP office is also coordinating with and leveraging investments made by other U.S. agencies (e.g., the DOE) as well as allies and partners (e.g., Canada and Australia). Beyond increased funding, the additional recommendations outlined in this chapter would help strengthen the DOD's ability to secure the critical minerals supply chain.

CHAPTER 9

An Evaluation of the Minerals Security Partnership

By Jane Nakano

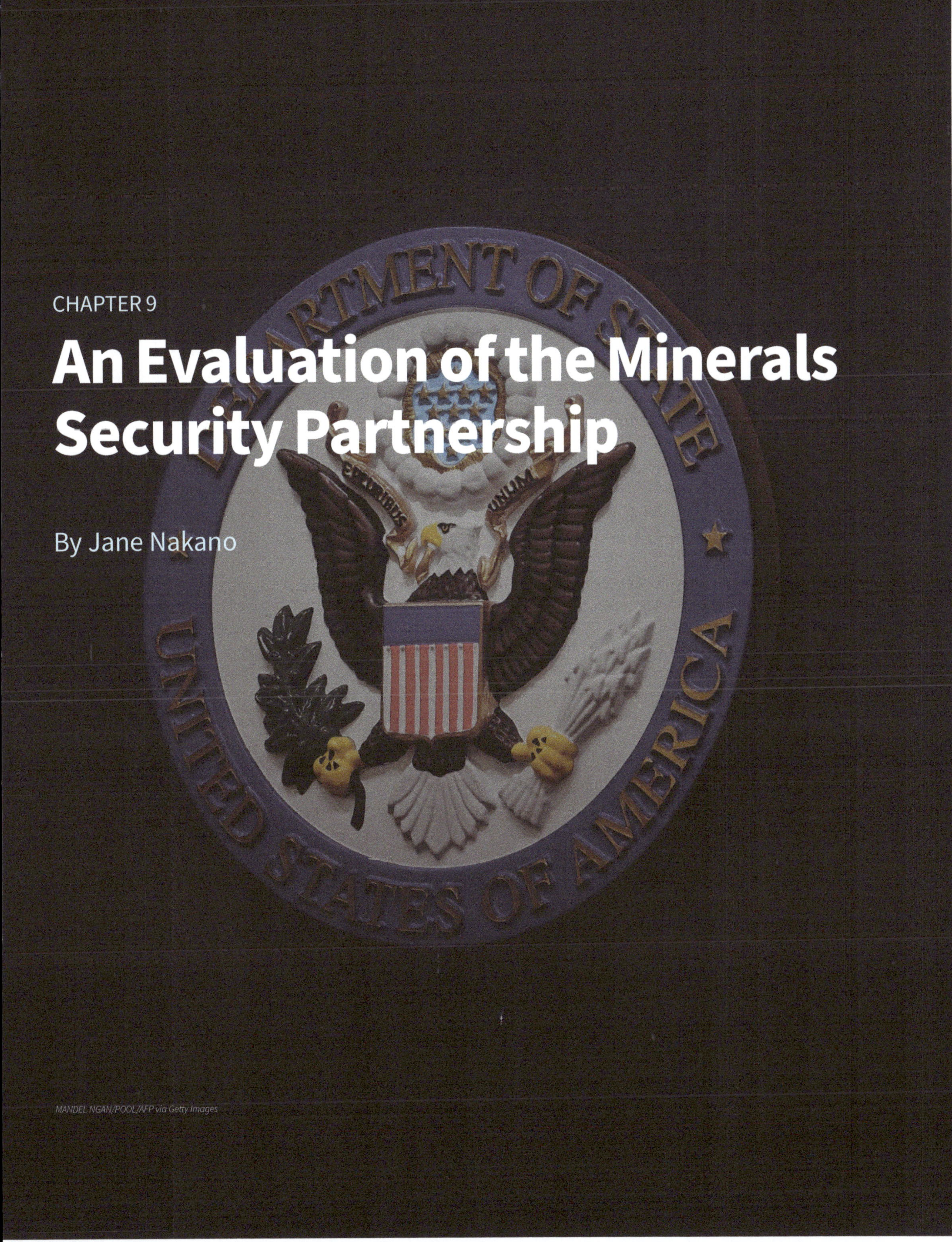

MANDEL NGAN/POOL/AFP via Getty Images

The MSP has successfully seeded the importance of mineral security in the global resource discourse.

International cooperation is one of the key pillars of the U.S. government strategy to help address vulnerabilities in the nation's supply chains for critical minerals.[266] In this context, the Minerals Security Partnership (MSP) is one major diplomatic initiative that specifically focuses on this issue. The MSP has successfully seeded the importance of mineral security in the global resource discourse, elevated critical minerals as a focus of U.S. resource diplomacy, and mobilized a coalition of market democracies to diversify global mineral supply chains. With several modifications, the MSP—as well as U.S. leadership through it—has the potential for a more dynamic, long-term, and durable presence.

MEMBERSHIP AND OBJECTIVES

Announced at the Prospectors and Developers Association of Canada (PDAC) convention in Toronto in June 2022, the MSP has seen its number of partners grow from 10 countries plus the European Union to 14 countries plus the European Union in the two years since its inception.[267] Some partners are global leaders in advanced technology manufacturing that are highly import dependent on component minerals, while others are mineral-rich advanced economies. A common thread appears to be the desire to see the global supply chains for critical minerals become more diversified and resilient to disruptions.

The MSP focuses on the full value chains—from mining to recycling—for minerals and metals that are "most relevant for clean energy technologies," such as lithium, cobalt, nickel, manganese, graphite, rare earth elements (REEs), and copper. In its own words, the MSP addresses four areas of major critical minerals challenges:

1. Diversifying and stabilizing global supply chains;
2. Investment in those supply chains;
3. Promoting high environmental, social, and governance standards in the mining, processing, and recycling sectors; and
4. Increasing recycling of critical minerals.[268]

In engaging mineral-rich third countries, the MSP is committed to ensuring that "minerals are produced, processed, and recycled in a way that helps countries realize the full economic benefits of their resources."[269] In fact, environmental, social, and governance (ESG) considerations are the defining value proposition to mineral-rich countries, as exemplified by repeated emphasis on these issues in press releases and remarks by MSP leadership.[270] The partnership positions itself as a global advocate for high ESG standards and a promoter of benefiting host localities. As such, resource-rich countries seeking MSP support are expected to demonstrate transparency around bidding processes and the engagement of local communities, while MSP partner countries and their firms are expected to adhere to high ESG standards.[271]

The initiative has emphasized the role of private investment and the importance of transparency in making global mineral supply chains more diversified, robust, and resilient. While the global mining industry has a few leading companies (commonly known as "mining majors") from market economies—such as Australia, the European Union, and the United Kingdom—the global supply chains for minerals for the energy transition are currently dominated by Chinese companies, whose ascent was facilitated by government support.[272] MSP leaders likely judged—correctly—that reducing China's preeminence would require private capital to augment the budgetary stream from its partner countries.

Prospective projects, submitted by MSP partners and driven by their companies, are not geographically limited to partner jurisdictions. Proposals are screened and then further evaluated by the project working group based on the investment climate, the need for the mineral, and the potential for investment.[273] The MSP supports selected projects by amplifying their efforts to attract investment and secure financing in various forms, including by offering political support, technical guidance, loans, political risk insurance, or financing through export promotion banks or export credit agencies.

MAJOR DEVELOPMENTS AND ACHIEVEMENTS

As of September 2024, the MSP has begun supporting about 30 projects: 16 on upstream mining and mineral extraction, 7 on midstream processing, and 7 on recycling and recovery.[274] These projects collectively source cobalt, copper, gallium, germanium, graphite, lithium, manganese, nickel, and REEs from around the world, including at least 13 projects in Africa, 6 in the Americas, 5 in Europe, and 3 in the Asia-Pacific.[275]

No single, official repository of MSP projects—including project terms and progresses—seems to exist in the public domain, but support has come in various forms. U.S. contributions through the MSP have included a $3.4 million technical assistance grant by the U.S. International Development Finance Corporation (DFC) for Pensana Rare Earths in Angola, a nonbinding DFC letter expressing potential loan provisions for Kabanga Nickel in Tanzania, and a nonbinding letter of interest by the Export-Import Bank of the United States (U.S. EXIM) for potential debt financing of up to $600 million toward the REE-focused Dubbo Project in Australia.[276] Other MSP partner support has included the arrangement of a $105 million debt facility by the German KfW IPEX-Bank toward a graphite project in Tanzania, as well as a mineral-related coordination framework between Congolese commodity-trading and mining company La Générale des Carrières et des Mines (Gécamines) and the government-affiliated Japan Organization for Metals and Energy Security (JOGMEC) that aims to advance coordination in mineral exploration, production, and processing.[277]

A major institutional development was the April 2024 establishment of the MSP Forum, which was set up to "deepen and cement the MSP's partnership with mineral-producing countries." Cochaired by the United States and European Union, the MSP Forum has two work streams, one focusing on project development and the other on policy dialogues.[278] Operationally, the project component will be led by the United States, and its scope "may include project information sharing among MSP partners, Forum members, and the private sector," as well as technical collaboration. The policy dialogue component, led by the European Union, might cover sustainable production, boosting local capacities, regulatory cooperation, application of high ESG standards, and effective recycling.[279]

Essentially, the forum has expanded the MSP's operational scope beyond identifying and supporting projects—likely in an effort to dispel the perception that MSP engagement is a new form of resource colonialism by advanced industrialized economies to exploit mineral-rich countries to fuel their industrial needs. While the forum likely helps increase the quality of communications between the MSP partners and forum members (i.e., mineral-rich, non-partner countries), membership does not seem to afford access—automatic or direct—to financial or technical assistance from the MSP, including access to benefits under the provisions of the U.S. Inflation Reduction Act. It remains to be seen whether the forum will effectively advance confidence building between the MSP partner countries and MSP Forum members.

More recently, in September 2024, development finance institutions and export credit agencies of the MSP partner governments established the MSP Finance Network. Aiming to "strengthen cooperation and promote information exchange and co-financing," the network includes nearly 30 participating institutions, including the DFC and U.S. EXIM.[280] This long-warranted development could enhance the scope and depth of MSP support, especially through the successful facilitation of cofinancing by multiple public and private institutions from MSP partner countries.

CHALLENGES IN ADVANCING MSP GOALS

While the MSP has furthered collective efforts to diversify global supply chains for minerals and metals that are important for the energy transition, several challenges have become evident.

Clarity of Purpose

Putting forward ESG as the MSP's preeminent value proposition likely stems from the desire of this U.S.-led group of advanced industrialized democracies to distinguish its global minerals activities from those of China. In fact, Chinese-owned mining projects in the Global South have increasingly become subject to ESG scrutiny, as civil society has reported and analyzed.[281] For example, the Business and Human Rights Resource Centre, a corporate watchdog, identified 102 alleged abuses—both human rights and environmental violations linked to Chinese companies' mining operations overseas, spanning 18 countries—from January 2021 to December 2022.[282]

However, even though the MSP seeks to actively "advocate" for high ESG standards—which form its core principles and shape the key criteria in mineral project selection, including the vetting of host countries and investor companies—the MSP has limited operational latitude on this front. The partnership is not set up to formulate its own set of standards or enforce a single set of standards for minerals production, processing, or recycling.[283] Instead, it defers to each partner country to apply and enforce its own standards, hoping to strike a delicate balance between respecting the judicial sovereignty of its individual partners and advancing ESG goals. While such pragmatism likely accords flexibility to MSP operations and agility to its inner workings, it may be obscuring the precise value proposition of the MSP.

It is difficult to ascertain whether—and to what extent—the application of high ESG standards might have limited the number of MSP projects approved or slowed the pace of project selection. Meeting ESG goals entails a long-term commitment that has not come easily even in advanced industrialized democracies. For example, the 2010 Dodd-Frank Wall Street Reform and Consumer Protection Act set the stage for financial disclosure under the auspices of the Securities and Exchange Commission (SEC) related to conflict minerals. However, the SEC has stopped enforcing actions against noncompliant companies as of 2017.[284] Therefore, the MSP might reassess the degree to which the application of high ESG standards should dictate its activities. The combination of the robust growth outlook for critical minerals demand and the extreme geographical concentration of mining and processing capacities generates an unprecedented sense of urgency in diversifying and expanding mineral supplies and supply chains. While few would likely dispute the importance of ESG concerns, there appears to be merit in recalibrating how ESG standards interact with the MSP's objective of "diversifying and stabilizing global supply chains" for critical minerals.[285]

Financial Support

The other major challenge for U.S. engagement with the MSP concerns financing. The primary tool for the U.S. government to unlock private capital for development needs is the DFC, which was created through the Better Utilization of Investments Leading to Development (BUILD) Act of 2018.[286] The DFC can use direct loans and loan guarantees, political risk insurance and reinsurance, equity investment, feasibility studies, and technical assistance to support private investments in low- and lower-middle-income countries.[287] It can also support activities in upper-middle-income countries "if such support is certified to have U.S. economic or foreign policy interests at stake and is designed for development impact."[288]

However, there is a misalignment between where the mineral resources are and the income classification of the host countries. Many prospective projects that could unlock additional supplies of critical minerals are in countries that are not in low- or lower-middle-income categories, including Australia (which has 28 percent of the global reserves of lithium and 21 percent of nickel), Brazil (which has 22 percent of graphite and 16 percent of nickel), Chile (which has 36 percent of lithium), and Indonesia (which has 21 percent of nickel).[289] Yet both Brazil and Indonesia are upper-middle-income countries, while Australia and Chile are high-income countries, based on World Bank classifications.

The DFC has a variety of tools to facilitate private investments in not only minerals production projects, but also infrastructure projects (e.g., roads and ports) that support the extractive sector (e.g., mining, quarrying, and oil and gas extraction). However, DFC support for extractive projects has been extremely limited. Only 14 of the 1,345 projects supported by the DFC and its predecessor agencies since 1999 are in the extractive sector. The pace is picking up, however. Nine of these fourteen projects have been approved since 2020, including six that are related to minerals and metals.[290]

The type of support for the six projects varies from technical assistance to equity investments, but the equity investment has been limited to two projects to date. The $105 million of equity is ring-fenced to two of TechMet's projects—an REE recovery project in South Africa and a nickel and cobalt project in Brazil.[291] The DFC has given a $150 million loan to Twigg Exploration and Mining for graphite mining and processing operations in Mozambique.[292] If the United States is to lead in strengthening and diversifying the global supply chains for critical minerals away from China, greater DFC support for minerals projects will be essential.

RECOMMENDATIONS

The MSP is a major multilateral platform in which the U.S. government has invested a considerable amount of diplomatic capital. Several areas are ripe for improvement and modifications, however, and the United States should pursue the following actions:

- **Extend the MSP's scope to include technical assistance.** The MSP plays a valuable coordinating role by bringing together key stakeholders and helping to identify and support minerals projects. One area in which the partnership could expand its support is technical assistance to MSP Forum members, especially in the form of geological surveys and resource mapping. Many mineral resources remain unmapped or under-mapped. For example, the absence of a systematic geological survey is one impediment to unlocking Africa's mineral resource potential.[293] In contrast to surveys conducted by an individual developer, even if it were from an MSP partner country, such mapping could become a communal asset that helps unlock private capital and could aid junior mining companies whose financial resources are more limited.
- **Better align DFC parameters with MSP objectives.** The following adjustments and modifications would make the DFC better equipped to aid the United States' role in the MSP, in turn empowering the partnership and granting it more operational flexibility.
  - **Remove the certification requirement for upper-middle-income countries for projects that contribute to critical minerals value chains.** Critical minerals projects, ranging from extraction to recycling, have "U.S. economic or foreign policy interests" by nature, given the importance and urgency of supply chain diversification.[294] This modification would not negate the importance of screening proposed projects for compliance with other DFC policies but could help redress the existing misalignment between where needed minerals are and the income classification of host countries.
  - **Relax the restriction against support to a majority state-owned or state-controlled entity.** DFC financing is presently geared toward private sector endeavors in the developing world.[295] However, restricting this support to companies that are mostly privately owned or controlled hinders the DFC's ability to engage key entities in the mining sector. For example, Gécamines, whose commanding ownership of cobalt and copper in Africa has made its coordination agreement with JOGMEC a highlight of recent MSP endeavors, is a state-controlled mining company. In some undertapped mineral-rich countries, the resources are wholly owned or controlled by the state. While reasons vary depending on the country and mineral, the state ownership of minerals can be a result of the process of nationalization that occurred in connection with the decolonialization of developing countries.[296] Permitting DFC support for state-owned entities could help enhance the institution's operational reach in a sector where the distinction between public and private may not be highly relevant—especially given the historical complexity regarding mineral ownership and the very notion of equity that the MSP Forum seeks to advance.
- **Begin considering a state financing tool that is not tied to an overseas development mandate.** While the above modifications merit serious consideration as Congress focuses on DFC reauthorization in 2025, it is also worth exploring if they merit establishing an institutional capacity similar to that of JOGMEC. Affiliated with Japan's Ministry of Economy, Trade, and

Industry (METI), JOGMEC has the authority to make strategic investments abroad to enhance Japan's energy security.[297] Free of restrictions on state ownership or control and income classification, JOGMEC support in the minerals sector around the world has included subsidies to 54 exploration projects, investment and loans to 23 exploration projects (including joint interest acquisition with Japanese companies), and loan guarantees to 8 mine development projects.[298] The most notable accomplishment is JOGMEC's strategic investment in Australia—a high-income country outside the DFC support scope—in the aftermath of the 2010 Chinese embargo on REE exports to Japan. The $250 million loan and equity deal to Lynas helped diversify Japan's REE supply sources while also saving the company from bankruptcy.[299] Lynas now produces 12 percent of the world's rare earth oxides and meets one-third of Japan's overall REE supply needs.[300]

CONCLUSION

The Minerals Security Partnership illustrates the collective desire of partner countries to counter China's dominance of global mineral supply chains. Whether the MSP can successfully turn this desire into collective support for economically viable and politically durable projects has significant implications for the pace of diversifying global supply chains and strengthening U.S. mineral security. Modifications to the MSP and DFC could put a new U.S. administration in a better position to succeed in addressing these pressing challenges.

SECTION 3

Addressing Challenges and Outstanding Questions in the Critical Minerals Industry

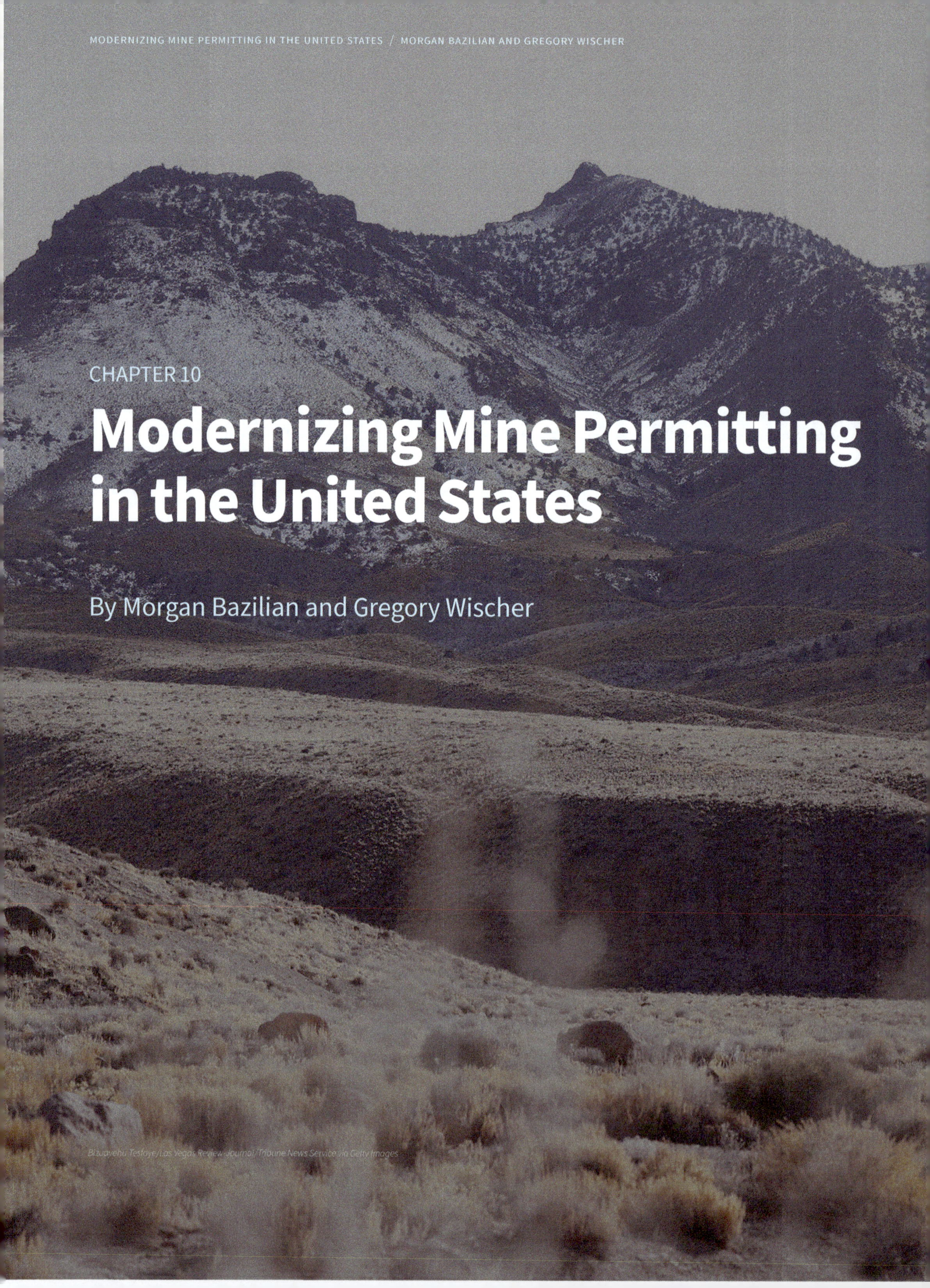

CHAPTER 10

Modernizing Mine Permitting in the United States

By Morgan Bazilian and Gregory Wischer

Bizuayehu Tesfaye/Las Vegas Review-Journal/Tribune News Service via Getty Images

In recent years, both Democratic and Republican presidents and bipartisan members of Congress have increasingly expressed the need to streamline permitting for new mine projects that are environmentally and socially responsible.

With domestic mineral demand forecasted to soar due to America's burgeoning re-industrialization and overseas mineral supplies imperiled by jurisdictional and shipping risks, members of the U.S. executive branch and Congress increasingly support a modernized permitting system that facilitates the development of domestic mining projects. They also generally back high permitting standards for safety, health, labor, emissions, and the environment, as well as Tribal consultation and community engagement. This emerging bipartisan consensus presents an opportunity for federal agencies to update rules and for Congress to pass laws streamlining permitting for new mines that are environmentally and socially responsible.

This chapter addresses the federal permitting process for new U.S. mines that extract hardrock minerals such as copper and nickel. It focuses predominantly on permitting for mine development—that is, building the mine—rather than exploration, production, or reclamation because permitting for mine development is more extensive than permitting for exploration and must also consider production and reclamation. This chapter first provides a historical overview of mine permitting in the United States before describing the current system of mine permitting. The final section recommends actions to the president and Congress to streamline permitting for domestic sustainable mining.

HISTORICAL OVERVIEW

The history of mine permitting in the United States can be categorized into three parts: mining expansion, mining efficiency, and environmental protection. For nearly its first hundred years of existence, the U.S. government—both the executive branch and Congress—deferred on establishing a general mining law or code.[301] U.S. courts mainly applied English laws and decisions when mining issues emerged, which rarely occurred.[302] However, the U.S. government's acquisition of mineral-rich western territories and the resulting legal questions necessitated concrete U.S. laws and regulations.[303] The U.S. government initially sought to generate revenue by leasing certain lands for mineral development, including lead and copper mines, but the system largely failed.[304]

Then in 1866, the federal government enacted a law declaring that federal public lands with minerals are "free and open to exploration" by U.S. citizens and those intending to become citizens, in alignment with "the local customs or rules of miners in the several mining districts."[305] Amendments to this law were enacted in 1870, and the Mining Act of 1872 ultimately clarified the process for individuals to explore public lands and, upon making a mineral discovery, to acquire mineral rights and develop mineral claims.[306] Given the prevailing national interest to incentivize mineral development, there were low federal fees and no federal royalties for mining on these lands.[307]

Thus, English common law usually applied to mineral development on the private lands east of the Missouri River, while the mining laws of 1866, 1870, and 1872 generally pertained to mineral development in the public lands west of the river.[308] These mining laws indeed proved successful in expanding U.S. mineral production.[309] (See the timeline at end of the chapter for a record of major U.S. federal laws affecting hardrock mining.)

The U.S. government tied the national welfare to the continued productivity of the mining industry, so amid declining ore grades in the 1910s, its attitude toward mining shifted to encompass not only development but also efficiency—that is, minimizing waste in the mining process.[310] As long as mining companies sought to reduce wasteful mining, the government generally urged a permissive view toward their activities.[311] It even prioritized efficiency during World War I, largely maintaining its system for permitting mines on public lands.[312] Still, the U.S. government lacked a national mining policy that applied to both public and non-public lands.[313]

Government attitudes into the mid-1930s continued to emphasize mining efficiency.[314] With the expansion of military programs and eventually the outbreak of World War II, the U.S. government sought to increase its mineral supply through federal mineral stockpiles and subsidies to mineral projects rather than permitting actions for the development of new mines.[315] Meanwhile, it continued to urge mining efficiency and minimizing waste.[316] The most notable permitting action in the 1940s was Reorganization Plan No. 3 of 1946, which authorized hardrock mineral leasing on acquired public lands and was anchored by prior laws like the Act of March 4, 1917.[317] During the Korean War, the U.S. government adopted similar stockpiling, subsidy, and mining efficiency policies as it did during World War II.[318]

However, some mines developed before and during this period had serious negative impacts on human health and the environment, including releases of pollutants that contaminated groundwater and surface water.[319] Starting in the late 1950s, members of Congress increasingly sought to pass laws to help protect the environment.[320] Many of these bills aimed to balance mineral development with environmental protections.[321] In 1969, Congress passed one of its most consequential pieces of environmental legislation, the National Environmental Policy Act (NEPA), and it was signed into law in 1970.[322]

At the time of the law's passage, the most noted part of NEPA was the creation of the Council on Environmental Quality (CEQ), which was tasked with creating new environmental programs and directing federal agencies to assess the environmental effects of their own programs.[323] But the most significant part of NEPA would arguably become Section 102. This section broadly outlines the NEPA administration process for federal agencies, including environmental impact statements (EISs) for "major Federal actions significantly affecting the quality of the human environment."[324] Major federal actions ultimately came to encompass federal agencies issuing certain permits.

After the enactment of NEPA, U.S. mine development slowed amid higher costs associated with environmental compliance, as well as higher energy, capital, and labor costs. In 1970, U.S. mineral exploration and development work declined for the first time since 1966, and mineral exploration and development work declined again the following year.[325] The U.S. Bureau of Mines noted at the time that health, safety, and environmental requirements "strongly influenced" U.S. mineral development.[326] In 1972, U.S. exploration and development work declined for the third-straight year amid the growing costs of environmental regulations and health and safety standards.[327] Congress that year passed several additional laws concerning the environment, including the Federal Water Pollution Control Act Amendments, which largely created the modern Clean Water Act.[328]

Mineral exploration and development work declined further in 1973, and additional environmental regulations contributed to higher costs for U.S. mines through the rest of the 1970s.[329] Capital expenditures per ton of output in the mining industry increased from $25–$30 per ton in the early 1970s to $75–$90 per ton in 1976.[330] Many new mines and capacity expansions were delayed or canceled.[331] Mine development not only became costlier but also took longer: By some estimates, the development timeline for U.S. mineral projects in 1976 was 15 years.[332] For permitting alone, a U.S. Forest Service (USFS) official estimated that the minimum processing time for a mineral lease and mining plan was 3 years.[333] Concerns thus arose in the 1970s that the NEPA process took too long.[334]

These concerns persist regarding the cost and length of the federal permitting process for new mines. As discussed in the next section, the laws and regulations that predominantly affect modern U.S. mine permitting today stem from laws and corresponding regulations enacted in the 1960s and 1970s.[335]

THE CURRENT SYSTEM

Developing a mine in the United States today requires federal, state, and local permits, authorizations, and compliance. Which permits are needed depends on the mine's location and plan of operations.[336] Mines located on public and non-public lands largely require the same federal permits, and a mine's plan of operations affects which specific permits are required based on how the mine might impact health, safety, and the environment (see Table 1). For example, an underground mine that plans to backfill with mine tailings would require an Underground Injection Control permit under the Safe Drinking Water Act.[337]

All mines require permits under federal programs, but the federal government has delegated permitting authority to the state governments for issuing certain permits.[338] For instance, most state governments issue National Pollutant Discharge Elimination System (NPDES) permits pertaining to Section 402 of the Clean Water Act.[339] Mines usually require NPDES permits because their plans of operations affect discharges for stormwater (e.g., rainwater) and mine contact water (e.g., runoff).[340] The Environmental Protection Agency (EPA) still retains oversight of state NPDES programs.[341] Similarly, most state governments have permitting authority over Clean Air Act permits, yet the EPA again maintains oversight.[342] These permits are often required for mines due to the crushing and dust associated with their operations.[343]

Table 1: Federal Permitting Requirements for Most U.S. Hardrock Mines

Permit/Compliance/Authorization Relevant Authority	Relevant Authority	Permit Rationale
Clean Water Act - National Pollution Discharge Elimination System (NPDES) permits	State governments, except in Massachusetts, New Hampshire, New affects discharges of stormwater like Mexico, and the District of Columbia; Environmental Protection Agency exercises oversight	Required if the mine's plan of operations affects discharges of stormwater (e.g., rainwater) and mine contact water (e.g., runoff)
Clean Water Act—Section 404 permit	Army Corps of Engineers, except in Michigan and New Jersey where state governments administer the program; Environmental Protection Agency exercises oversight	Required if the mine's plan of operation involves dredging or filling material in locations considered jurisdictional waters like adjacent wetlands-covered by the Clean Water Act
Clean Water Act Section—401 certification	State governments; Environmental Protection Agency exercises oversight	Required to issue a Section 402 permit issued by the Environmental Protection Agency and a Section 404 permit issued by the US Army Corps of Engineers
National Historic Preservation Act Section 106 compliance	State Historic Preservation Officer or Tribal Historic Preservation Officer	Required to issue a Section 404 permit
Endangered Species Act—Section 7 US Fish & Wildlife Service compliance	US Fish & Wildlife Service	Required to issue a Section 404 permit
National Environmental Policy Act—Record of Decision and Plan of Operations approval	• Non-public lands: US Army Corps Required if the mine requires a Section • National Forest System lands: US Forest Service • Other public lands: Bureau of Land Management *Other federal agencies may act as 'cooperating agencies" in the NEPA	Required if a mine requires a 404 permit or is located on federal lands
Safe Drinking Water Act Underground Injection Control (UIC) permits	Depending on the well class, either the Environmental Protection Agency or state government	Required if the mine's plan of operations involves underground disposal of wastewater, underground mine tailings backfill, in-situ recovery operations, or similar activities
Explosives permit	Bureau of Alcohol, Tobacco, Firearms and Explosives	Required for transporting, storing, and using explosives at the mine
Radio authorization	Federal Communications Commission	Required for installing and operating radio systems at the mine
• Mine identification number • Legal Identification Report • Part 48 training plan	Mine Safety and Health Administration	Required before commencing mining operations

Source: Author's analysis.

As for permits needed directly from the federal government, mines generally need a Section 404 permit under the Clean Water Act.[344] This permit is necessary if the mine's plan of operations involves dredging or filling material—for example, installing a culvert for a stream crossing or building a storage facility for waste rock or tailings—in locations considered jurisdictional waters subject to the Clean Water Act.[345]

In all states except Michigan and New Jersey, the federal government via the U.S. Army Corps of Engineers (USACE) retains authority over issuing Section 404 permits.[346] The EPA maintains veto power over these permits in all states—and the EPA has exercised this veto, such as over the Pebble Mine project in Alaska.[347] For the USACE to issue a Section 404 permit, the mine project must also receive Clean Water Act Section 401 certification—which the state government issues—and comply with Section 7 of the Endangered Species Act and Section 106 of the National Historic Preservation Act.[348]

Importantly, when a federal agency decides to grant a permit like a section 404 permit or approve a mine on public lands, it is considered a major federal action under NEPA and thus requires the lead federal agency to prepare an EIS.[349] This process requires the lead federal agency to assess the environmental impacts of its proposed action and possible alternative actions, including no action.[350] After receiving the mine application, plan, and related technical reports, the lead agency scopes the EIS, authors the draft EIS, prepares the final EIS, and issues a Record of Decision, which is the federal government's final determination on whether to approve a project's plan of operations (see Figure 1).[351]

For a mine that seeks a section 404 permit on non-public lands, the USACE is the lead agency in the NEPA process. The EPA reviews and comments on the USACE's EISs and can act as a cooperating agency.[352] To illustrate the length of the process, the USACE's average timeline in 2021 across all types of projects, from publishing a notice of intent for drafting an EIS to publishing a final EIS, was 5.8 years.[353] The Fiscal Responsibility Act of 2023 did establish a two-year timeline for the federal government to complete EISs, but the lead federal agency can extend the deadline.[354]

The Tamarack Nickel project in Minnesota offers an example of this permitting process. Even though it is located wholly on non-public lands, including both state lands and private lands, federal permits are likely required, including a Section 404 permit under the Clean Water Act and, thus, Section 106 compliance under the National Historic Preservation Act and Section 7 compliance under the Endangered Species Act.[355] Consequently, the project must follow the NEPA process, with the USACE expected as the lead agency.[356]

For hardrock mines on public lands, the lead federal agencies for NEPA reviews are either the Bureau of Land Management (BLM) within the Department of the Interior or the USFS within the Department of Agriculture.[357] While the BLM administers mining claims on all public lands, including regulations for preventing "unnecessary or undue degradation" to public lands, the relevant surface management agency—either the BLM or the USFS—oversees the mine development.[358] USFS regulations for mining are practically unaltered since 1974, and BLM regulations are largely unchanged since 2001.[359]

Generally, the BLM and USFS permit mines on public lands in the process depicted in Figure 2.[360] For all types of projects reviewed by the USFS in 2021, the average timeline from publishing a notice of intent for drafting

Figure 1: Overview of National Environmental Policy Act's Environmental Impact Statement Process

Source: Authors elaboration based on Environmental Protection Agency, "Mining Issues: EPA Region 9 Regional Tribal Operations Committee Meeting," presentation, April 20, 2023, slide 18, https://www.epa.gov/system/files/documents/2023-04/r9-rtoc-presentation-mining-issues-breakout-spring-2023.pdf.

Figure 2: Overview of the Bureau of Land Management and U.S. Forest Service's Process to Approve Hardrock Mine Plans

Source: Authors elaboration based on U.S. Government Accountability Office (GAO), Hardrock Mining: BLM and Forest Service Have Taken Some Actions to Expedite the Mine Plan Review Process but Could Do More, GAO-16-165 (Washington, DC: GAO, January 2016), 10–11, https://www.gao.gov/assets/gao-16-165.pdf.

an EIS to publishing a final EIS was 5.8 years.[361] For new hardrock mining projects approved by the BLM between FY 2013 and April 2024, the average timeline for the entire process—from the project appearing in the BLM's records to the BLM authorizing ground-disturbing activities—was 4.6 years.[362] Thus, regardless of the lead federal agency, the NEPA process can be expected to take around five years.

Under the Mining Act of 1872, the BLM and USFS further regulate mining on public lands based on whether the public lands are non-acquired or acquired.[363] Non-acquired public lands have always been federally owned, while acquired lands have been obtained by the federal government, such as through a purchase, gift, or condemnation.[364]

On non-acquired public lands, the location system applies to hardrock minerals.[365] After receiving authorization, firms and individuals can mine these minerals if the public lands are not closed or withdrawn from mineral entry and by paying a one-time $49 location fee, a $25 processing fee per new claim, and an annual $200 maintenance fee.[366] Mineral production under the location system is not subject to a federal royalty, although some entities—including the Biden administration's Interagency Working Group on Mining Laws, Regulations, and Permitting—supported imposing federal royalties on these minerals.[367] Conversely, on acquired public lands, the leasing system applies to hardrock minerals.[368] If the public lands are open to mineral activity, firms and individuals can mine these minerals after receiving authorization, but the mineral production is subject to federal royalties.[369]

To illustrate the above permitting process, the Thacker Pass lithium project in Nevada is on BLM-managed lands.[370] Correspondingly, plans of operations and reclamation had to be submitted, and because approving the plan of operations was considered a major federal action, the BLM became the lead agency in the NEPA process.[371] Although the plan of operations was received in September 2019 and the final Record of Decision was reached in January 2021, the project faced legal challenges until March 2024 concerning its potential impact on properties spiritually, culturally, and historically significant to several Tribes.[372] Thus, even a relatively quick permitting process can be followed by lengthy litigation.[373]

In another example, the Idaho Cobalt Operations is on USFS land.[374] The USFS was, therefore, the lead agency in the NEPA process, with the EPA acting as a cooperating agency.[375] With the mine's plan of operations changing in 2006, it took eight years from the USFS publishing a notice of intent for drafting an EIS in 2001 to reaching a Record of Decision in 2009.[376] As with many mine projects, the greatest environmental concerns were the mine's potential impacts on surface water and groundwater.[377] To mitigate these risks, the approved plan of operations addressed factors such as tailings and waste rock management.[378]

Notably, certain types of mines can receive special permitting coverage under Title 41 of the Fixing America's Surface Transportation Act (FAST-41) of 2015. FAST-41 created the Permitting Council to streamline federal permitting by coordinating federal environmental permits and approvals for certain projects; it also created the Federal Permitting Dashboard to enhance permitting timeline transparency and predictability.[379] In 2021, mining was added as a sector eligible for FAST-41 coverage.[380] To be eligible, a mining project must meet different requirements depending on the specific criteria. For example, the "objective" criteria require that the project be subject to NEPA, need more than $200 million in investment, and not qualify for other expedited permitting processes.[381]

In May 2023, the Hermosa manganese-zinc project in Arizona became the first mining project covered by FAST-41.[382] Although the Hermosa project will involve extensive mine development and mineral extraction on private lands, for which it will need state permits, the planned expansion of the mine operations will include USFS land, invoking the NEPA process and making the project eligible for FAST-41 benefits under the objective criteria.[383] With FAST-41 coverage, the Hermosa project expects a more efficient permitting process, and it anticipates a Record of Decision in 2027.[384]

RECOMMENDATIONS FOR STREAMLINING THE PROCESS

The U.S. government—both the executive branch and Congress—could improve the process for permitting new domestic mines by improving permitting clarity, cost, and timelines. An inefficient permitting regime dissuades those seeking to build new mine projects and can cause delays or even cancellation for projects in development. Thus, an inefficient permitting regime limits the pipeline of mines, and those projects take a long time to come online, both constraining and delaying new mineral supply.

Streamlining permitting for environmentally and socially responsible new mines has garnered bipartisan federal support, including from both the Trump and Biden administrations.[385] For example, the Trump administration's Executive Order 13953 directed agency heads to, "as appropriate and consistent with applicable law, use all available authorities to accelerate the issuance of permits and the completion of projects in connection with expanding and protecting the domestic supply chain for minerals."[386] In Congress as well, members of both parties have sought to streamline permitting for mines, best evidenced by the Energy Permitting Reform Act of 2024, introduced by Senators Joe Manchin (I-WV) and John Barrasso (R-WY).[387]

The following recommendations seek to promote sustainable mining, helping the United States strengthen its national security, grow its economic prosperity, and pursue its environmental objectives. Recommendations are specifically tailored for the federal government—both the executive branch and Congress—not state or local governments.

RECOMMENDATIONS TO THE PRESIDENT

- **Issue a new rule that enhances Tribal consultation and community engagement with mine applicants in the NEPA process.**

  To better address the input of Tribes and local communities affected by permitting new mines, the next presidential administration could issue a new rule that enhances Tribal consultation and community engagement in the NEPA process.[388] An oft-cited study using 2021 data found that a majority of mine sites for nickel, copper, lithium, and cobalt in the United States are located within 35 miles of Native American reservations.[389] In the NEPA process, however, federal agencies—not the mining applicants—are required to consult with the Tribal authorities, who have noted that this interaction is often not timely or constructive.[390] This new rule should require mining companies to consult with the requisite Tribes and engage with the local communities according to set guidelines upon beginning the NEPA process, such as by conducting onsite tours, holding mandatory monthly meetings, and issuing automatic project notifications.[391] Such engagement and consultation would help address issues and thus avoid lawsuits that may arise concerning the NEPA process.

- **Expand FAST-41 coverage under the Permitting Council's "discretionary" criteria to mining projects that will extract energy transition minerals.**

  To increase the number of mining projects covered by FAST-41, the next president could direct the Permitting Council to exercise its voting authority under the discretionary criteria to cover more projects that will extract minerals on the Department of Energy's "critical materials for energy" list.[392] The only mining project covered by FAST-41 as of September 2024 is South32's Hermosa project, despite mining projects having become eligible for FAST-41 coverage in January

2021.[393] Under the objective criteria, a mining project like the Idaho Cobalt Operations would not have been eligible for FAST-41 coverage—even though the mine was subject to NEPA and did not qualify for other expedited permitting processes—because it required less than $200 million in investment.[394] The Permitting Council could bypass the objective criteria's total investment requirement by voting to cover specific mine projects that will extract energy transition minerals.

- **Issue a new rule for the CEQ's "intensity" factors that establishes clear thresholds for potentially significant impacts.**

  To help federal agencies ascertain if projects require EISs, the new administration's CEQ could issue a new rule for "intensity" factors that establishes clear thresholds for potentially significant environmental impacts. Determining if a time-consuming EIS is required instead of a simpler environmental assessment (EA)—which can take roughly half the time to complete—depends on whether a major federal action (e.g., granting a federal permit) may significantly impact the environment.[395] The lead agency in the NEPA process determines the significance based on the "context" and "intensity" of the action's impact, and the CEQ defines the context and lists 10 intensity factors.[396] A conclusion or "substantial question" that the action may have a significant impact for one factor triggers an EIS; however, the CEQ does not currently define thresholds for triggering each factor.[397] A new rule that clarifies these definitions would enable agencies to pursue EAs when appropriate rather than favoring caution by immediately drafting an EIS.

RECOMMENDATIONS TO CONGRESS

- **Increase funding for more agency staffing to support NEPA reviews.**

  To better ensure timely NEPA reviews, Congress could increase funding for agencies to hire more mineral experts.[398] In 2016, BLM and USFS officials reported that field offices, which are the specific agency units leading the NEPA process, have limited mining expertise, causing delays.[399] Additional staffing could enable agencies to engage in early planning meetings with mine project applicants and help reduce applicant-caused delays, such as submission of mine plans that lack adequate information and clarity.[400] Thus, adequate staffing with the necessary expertise could help facilitate the NEPA process.[401]

- **Increase funding for permit applicants to offset costs associated with the permitting process.**

  To help reduce the costs faced by mine applicants, Congress could appropriate more funding to defray permitting-related costs.[402] Indeed, the Department of Defense already offers and has disbursed Defense Production Act Title III grants to mine projects to help cover costs related to the NEPA process.[403] This financial support can be increased and expanded to cover additional costs. For instance, mine applicants generally have to pay for third-party contractors assisting in the NEPA process, yet the quality of these contractors varies and leads to corresponding delays.[404] Congress could pass funding for permit applicants to hire approved contractors with a track record of responsive, high-quality work in the NEPA process, both reducing mine applicants' compliance costs and streamlining the NEPA process.

- **Establish categorical exclusions (CEs) for the construction, expansion, or modernization of mines that will produce energy transition minerals.**

  To satisfy environmental requirements under NEPA while expediting permitting timelines, Congress could pass legislation—similar to the Building Chips in America Act of 2023—that treats the permitting of mines for energy transition minerals as categorically excluded from EA and EIS requirements.[405] Illustrating the faster permitting timeline for CEs, the USFS took an average of seven months to complete a CE between 2005 and 2020.[406] To be covered by the CE, a project must produce at least one energy transition mineral as the primary product. For example, an iron ore mine that produces neodymium as a byproduct would not be subject to a CE. Importantly, the covered energy transition minerals should be listed in the legislation and could include those minerals on the Department of Energy's critical energy materials list.[407]

Timeline of Major U.S. Federal Laws That Affect Hardrock Mining

1785

Land Ordinance of 1785: Reserved to the Congress—under the Articles of Confederation—for its own use one-third of all gold, silver, lead, and copper mines in the lands west of the Appalachian Mountains, north of the Ohio River, and east of the Mississippi River

1807

Act of March 3, 1807: ("An act making provision for the disposal of the public lands, situated between the United States military tract and the Connecticut reserve, and for other purposes"): Authorized the president to lease lead-bearing lands in the Indiana Territory

1846

Act of July 11, 1846: ("An act to authorize the President of the United States to sell the reserved mineral lands in the states of Illinois and Arkansas, and territories of Wisconsin and Iowa, supposed to contain lead ore"): Authorized the president to sell the lead-bearing lands in Illinois, Arkansas, and the territories of Wisconsin and Iowa

1847

Act of March 1, 1847: ("An act to establish a Land Office in the northern part of Michigan, and to provide for the sale of mineral lands in the state of Michigan"): Authorized the president to sell the lands containing copper, lead, and other valuable ores in Michigan

1849

Act of March 3, 1849: ("An act to establish the Home Department, and to provide for the Treasury Department an Assistant Secretary of the Treasury, and a Commissioner of Customs"): Transferred the powers exercised by the secretary of the treasury over lead mines and other mines to the secretary of the interior

1866

Act of July 26, 1866: ("An act granting the right of way to ditch and canal owners over the public lands, and for other purposes"): Opened mineral exploration and development on public lands to U.S. citizens and those intending to become citizens, and established corresponding procedures for the ownership of vein and lode mineral deposits

1870

Act of July 9, 1870 ("An act to amend 'An act granting the right of way to ditch and canal owners over the public lands, and for other purposes'"): Expanded the 1866 mining law to establish a procedure for the ownership of placer mineral deposits

1872

Mining Act of 1872: ("An act to promote the development of the mining resources of the United States"): Unified the 1866 and 1870 mining laws, and clarified the procedures and fees for mineral ownership on public lands

1891

Act of February 28, 1891: Amended the General Allotment Act of 1887, and authorized mineral leasing on certain Tribal lands

Act of March 8, 1891: (Forest Reserve Act of 1891): Authorized the president to reserve public lands as national forest reserves

1897

Act of June 4, 1897: (Organic Act of 1897): Authorized the secretary of the interior to manage the use of national forest reserves, including making them available for mining

1899

Rivers and Harbors Appropriation Act of 1899: Prohibited the erection of obstructions to navigable waters unless authorized by the federal government; the oldest federal environmental law in the United States

1910

Act of June 25, 1910 (Pickett Act): Authorized the president to temporarily withdraw public lands from settlement, location, sale, or entry and reserve them for public purposes

1916

Stock Raising Homestead Act of 1916: Maintained that the federal government owned the mineral rights on lands designated for "stock raising," enabling private persons to explore, own, and mine minerals on these lands even if homesteaders owned the surface rights

1917

Act of March 4, 1917 (Weeks Act Minerals): Authorized the secretary of agriculture to permit mineral development on lands acquired under the Weeks law

1919

Act of June 30, 1919: Authorized the secretary of the interior to issue leases for mining metalliferous minerals on unallotted Tribal lands in nine western states.

1938

Indian Mineral Leasing Act of 1938: Permitted mineral leasing on unallotted lands within Indian reservations, subject to the authorization of the requisite Tribe and approval by the secretary of the interior

1946

Reorganization Plan No. 3 of 1946: Established the leasing of hardrock minerals on acquired federal lands—which are lands granted or sold to the United States—and corresponding oversight by the secretary of the interior

1954

Multiple Mineral Development Act of 1954: Established procedures for the mineral development of lands subject to both the 1872 mining law and the Mineral Leasing Act of 1920

1955

Act of July 23, 1955 (Surface Resources Act): Clarified that surface rights could be managed separately from mineral rights, restricting surface activities on mining claims on public lands to activities "reasonably incident" to prospecting, mining, or processing

1963

Clean Air Act of 1963: Established air quality standards, leading to permit requirements for certain levels of mining-related dust, crushing and processing, and emissions from power generation equipment

1966

Federal Metal and Nonmetallic Mine Safety Act of 1966: Established inspection, safety, and health procedures for metal and nonmetal mining operations

National Historic Preservation Act of 1966: Established requirements for the federal government to consider the impact of its actions on historic, cultural, and spiritual sites

1969

National Environmental Policy Act of 1969: Established requirements for the federal government to assess the potential environmental impacts of its actions, such as the approval of mines on federal lands and issuance of certain federal permits

1970

Resource Recovery Act of 1970: Directed the federal government to establish guidelines for solid waste recovery, collection, separation, and disposal

Occupational Safety and Health Act of 1970: Established standards to protect workers from recognized workplace hazards

Mining and Mineral Policy of 1970: Declared the federal government's policy and national interest in developing an economically robust and environmentally responsible mining industry in the United States

1972

Federal Water Pollution Control Act Amendments of 1972 (framework for the modern Clean Water Act): Established standards for discharging pollutants—including stormwater and mine runoff—as well as dredged or fill material into waters of the United States

1973

Endangered Species Act of 1973: Established protections for endangered and threatened species and their habitats, subjecting mine projects to certain regulations during project development and operation

1974

Safe Drinking Water Act of 1974: Authorized the Environmental Protection Agency to protect underground sources of drinking water, resulting in requirements for Underground Injection Control permits for the underground disposal of wastewater, underground mine tailings backfill, and in situ recovery operations

1976

Federal Land Policy and Management Act of 1976: Established guidelines for the uses (e.g., mining) and management of public lands, including preventing "unnecessary or undue degradation of the lands"

Resource Conservation and Recovery Act of 1976: Authorized the Environmental Protection Agency to regulate the management of solid and hazardous wastes; subsequent regulations exempted mining waste rock and tailings from federal hazardous waste regulations

1977

Federal Mine Safety and Health Act of 1977: Combined and enhanced the health and safety standards for coal, metal, and nonmetal mines

1982

Indian Mineral Development Act of 1982 (Melcher Act): Authorized Tribes to enter into any form of agreement approved by the Secretary of the Interior for the development of mineral resources on Tribal lands and to sell mineral resources produced on Tribal lands

1992

National Historic Preservation Act Amendments of 1992: Established requirements for the federal government to consult with Tribes for actions that may affect sites of historical, cultural, or spiritual significance on or off Tribal lands; allowed Tribes to assume the role of state historic preservation officers as "Tribal historic preservation officers"

1993

Act of April 16, 1993 ("An act to amend the Stock Raising Homestead Act to resolve certain problems regarding subsurface estates, and for other purposes"): Enhanced the requirements for mineral exploration and development on stock raising homestead lands, including requiring written notice to and consent from the surface owner for mining-related activities

2006

Mine Improvement and New Emergency Response Act (MINER Act) of 2006: Amended the Federal Mine Safety and Health Act of 1977 by enhancing emergency preparedness, prescribing penalties for operator violations, and authorizing the secretary of labor to bring civil actions when an operator breaches orders or decisions under the act

2015

Fixing America's Surface Transportation (FAST) Act of 2015: Established a Permitting Council to improve the transparency and predictability for permitting certain infrastructure projects, including covered mining projects

CONCLUSION

For much of its history, Congress sought to impose a permitting regime that fostered and supported U.S. mineral development, given the importance of minerals to national security and economic prosperity. However, in the 1960s, the legislative branch became increasingly concerned about the impact of mining on human health and the environment. It successfully passed a series of laws to help address these issues, and the resulting regulations and judicial review from administering these laws still affect the permitting of new mines. In recent years, both Democratic and Republican presidents and bipartisan members of Congress have increasingly expressed the need to streamline permitting for new mine projects that are environmentally and socially responsible.

These officials cite the necessity of new mines for not only national security and economic reasons but also environmental sustainability, including the adoption of new energy technologies. With the public interest of the government and the private interest of the mining industry generally supportive of streamlined mine permitting, both the incoming Trump administration and Congress could modernize the permitting process for new mines in the United States. As U.S. Geological Survey director George Otis Smith wrote in 1919, "Public interest and private interest in the long run are less antagonistic than either the captain of industry or the public servant has suspected."[408] Public and private interests now align on the importance of permitting new mines while simultaneously upholding high environmental and social standards.

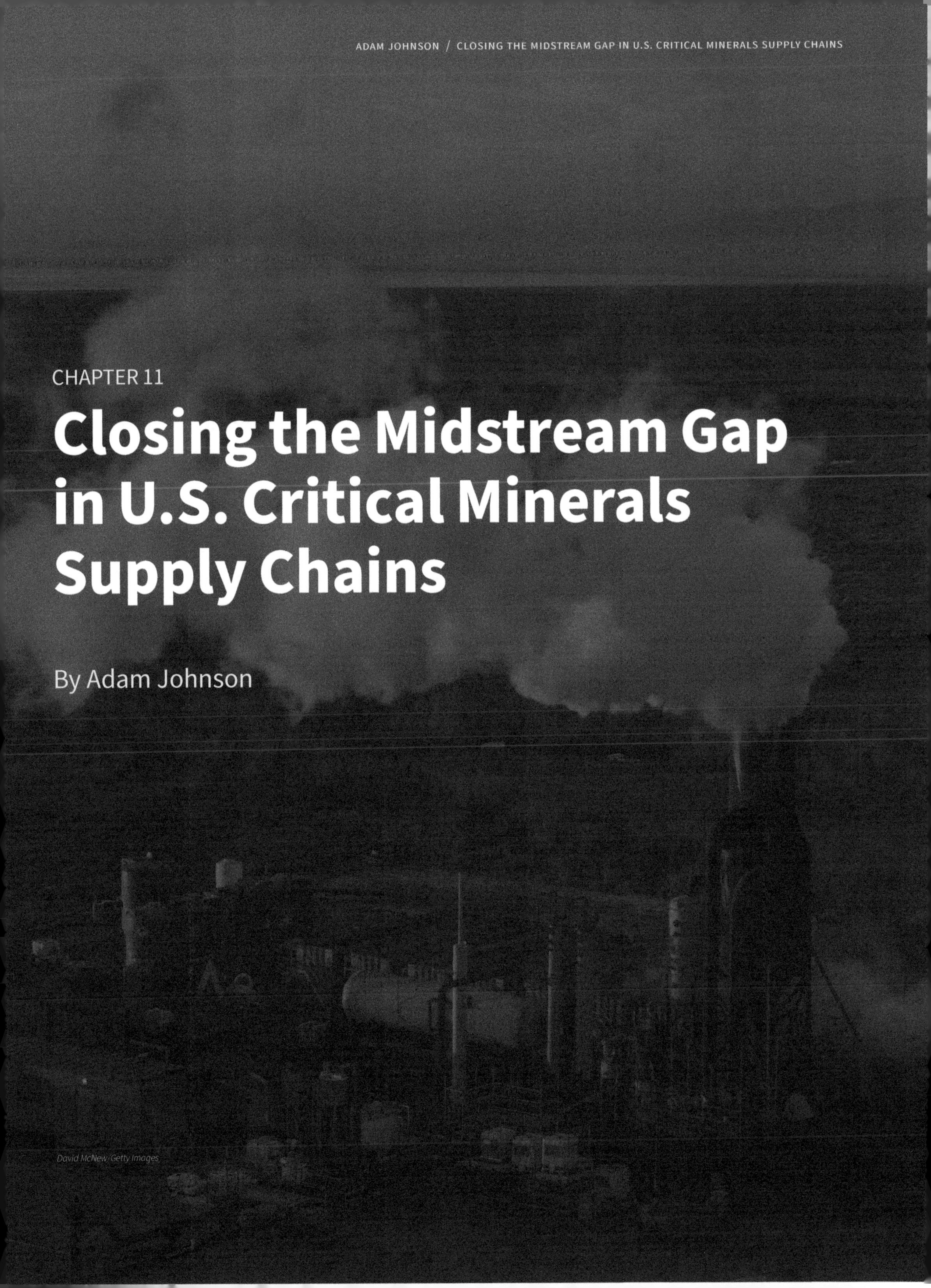

CHAPTER 11

Closing the Midstream Gap in U.S. Critical Minerals Supply Chains

By Adam Johnson

David McNew/Getty Images

Closing the midstream gap in the U.S. supply chain is crucial for safeguarding strategic interests and maintaining leadership in the global technologies that will shape the future.

Critical minerals have become essential to the technologies that power modern economies. While much attention is given to mining, it is the midstream stage—the processing and refining of these raw materials—that poses the greatest challenge for the United States. Processing these minerals into usable materials is a critical, yet often overlooked, part of the supply chain. It is essential to explore the key obstacles to developing strong midstream capabilities, examining the historical, economic, and geopolitical factors that have contributed to the current gap.

Although the United States has made progress in increasing its mining capacity, the country still struggles to efficiently process these critical minerals. Addressing this midstream bottleneck is essential for reducing dependence on foreign adversaries and building a more resilient supply chain.

THE MIDSTREAM BOTTLENECK

A Strategic Blind Spot

The U.S. critical minerals supply chain is missing a key component: midstream processing and the conversion of raw minerals into advanced materials. This stage involves transforming mined ores into high-purity metals and producing materials tailored for specific end uses. Although the United States is a significant producer of minerals such as copper and rare earth elements (REEs), it is a net exporter of both because it lacks the infrastructure needed to process them domestically. In 2023, the United States produced more than 12 percent of the REEs mined globally but exported 93 percent of those materials.[409] During the same period, the United States produced 5 percent of the world's mined copper and exported 32 percent of it.[410]

These important raw materials—whether mined or recycled—are considered critical to national defense and economic security, yet they are exported at high rates, often to China, where they are processed and converted into finished products. As a result, the United States finds itself exporting its raw materials only to buy them back in the form of advanced technologies like cell phones and electric vehicles. This reliance creates strategic risks that directly affect national security and economic autonomy.

Without robust midstream processing capabilities, the United States is exposed to serious potential supply chain disruptions. Geopolitical tensions, export restrictions, or price manipulation could significantly impact the availability of critical materials. In recent years, China has implemented multiple export restrictions on critical minerals, particularly targeting raw materials such as gallium, germanium, and REEs, which are crucial for advanced technology manufacturing. In July 2023, the Chinese Ministry of Commerce announced export controls on gallium and germanium products, which took effect on August 1, 2023.[411] These controls require companies to apply for special export licenses, effectively reducing global supply amid fears of a broader Chinese strategy aimed at countering Western semiconductor production capabilities. China has also shown a willingness to restrict other strategic resources, like REES, of which it refines 92 percent of the global supply.[412] Beijing has imposed export restrictions on rare earth processing and magnet technologies, underscoring its readiness to use its near-monopoly position as leverage in ongoing trade disputes, particularly with the United States and the European Union.[413]

History shows that these disruptions are likely to occur at the least opportune times, whether due to a global pandemic, such as Covid-19, or international conflicts, such as the war in Ukraine.

This lack of domestic processing capability means that some of the most crucial U.S. technologies remain vulnerable to the influence of foreign adversaries. The

Figure 1: Share of Top Three Processing Countries of Selected Minerals, 2022

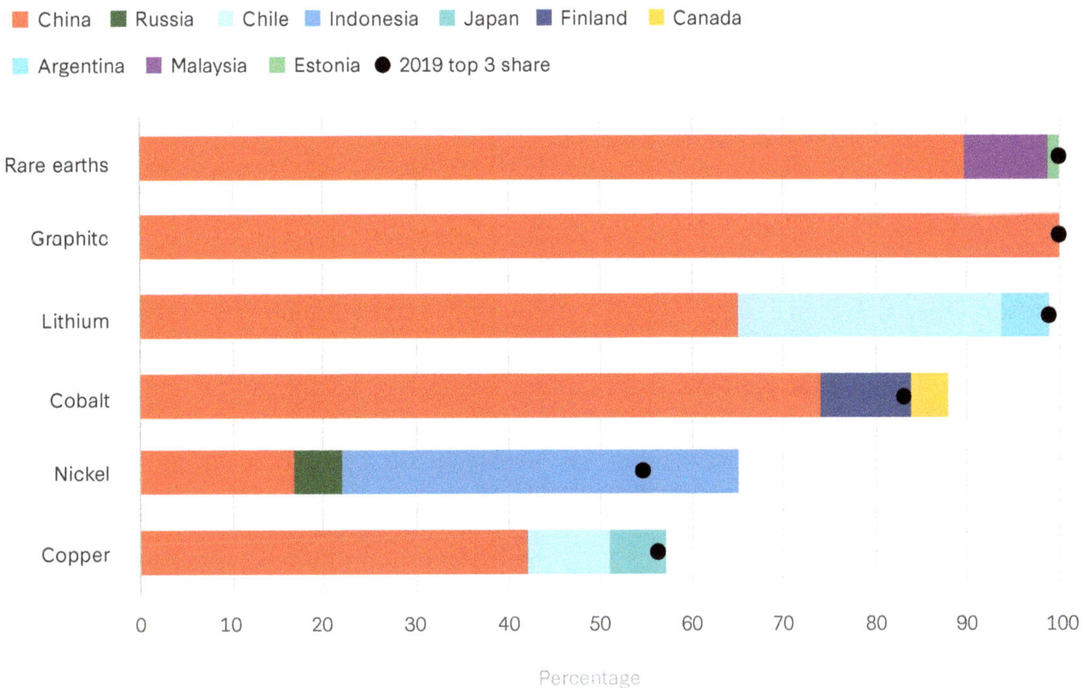

Source: This is a work derived by CSIS from IEA material. CSIS is solely liable for this derived work. The derived work is not endorsed by the IEA in any manner.

fact that China holds dominant positions in processing for many essential minerals, including lithium (65 percent), cobalt (74 percent), copper (42 percent), and graphite (100 percent), makes this dependence especially concerning.[414] As global demand for critical minerals rises, authoritarian regimes like those in China and Russia are exploiting the West's hesitancy toward mining and refining, allowing them to nurture U.S. dependence in a deliberate effort to weaken Western technological leadership and economic influence.

Closing the midstream gap in the U.S. supply chain is crucial for safeguarding strategic interests, reinforcing economic independence, and maintaining leadership in the global technologies that will shape the future. This challenge is not insurmountable, but it would be a mistake to underestimate the urgency or complexity of the task.

PROCESSING METHODS
Decoding the Complexities

Midstream processing of minerals typically begins with physical beneficiation, where the ore is crushed and the valuable minerals are separated from waste rock through techniques like flotation or magnetic separation. However, many minerals require further refining to reach a pure, usable form. At this stage, two primary methods are used: pyrometallurgy and hydrometallurgy.

Pyrometallurgy involves the use of high temperatures to extract metals, especially from ores that are rich in sulfides, such as nickel, cobalt, and copper. In this process, the ore is heated in a furnace, triggering chemical reactions that separate the metal from other elements. While this technique is well established and effective for certain ores, it is energy intensive and produces harmful emissions. The significant energy and environmental costs associated with pyrometallurgy make it less appealing for widespread use, especially as industries seek cleaner, more sustainable processing methods.

Hydrometallurgy, by contrast, uses chemicals to dissolve metals from ores, allowing them to be recovered from the solution. This method is often applied to oxide ores, such as those containing lithium or copper, as well as laterites, which are a key source of nickel and cobalt. Hydrometallurgy tends to be more energy efficient than pyrometallurgy and can have a lower environmental impact, provided the chemicals are managed responsibly. For example, different hydrometallurgical techniques are needed in lithium production depending on whether the lithium is extracted from spodumene ore (a solid rock) or brine (saltwater). Brine extraction traditionally involves evaporating water to concentrate the lithium, but newer technologies like direct lithium extraction (DLE) are being developed to speed up the process and reduce water usage, making it more sustainable.

Midstream processing of minerals often involves byproduct extraction, especially when dealing with minerals such as gallium or germanium, which are rarely found in high concentrations on their own. For example, gallium is primarily obtained as a byproduct of aluminum production from bauxite, while germanium is recovered from zinc ores. Extracting these byproducts involves hydrometallurgical processes that dissolve the primary metal-bearing minerals, from which trace elements can be separated. Gallium is extracted from the caustic liquor used in the Bayer process for refining bauxite into alumina, which contains only small parts per million of gallium. The gallium is precipitated through additional chemical treatments, making its production more complex and economically viable only when bauxite production volumes are high and market prices justify the added expense.

Similarly, germanium is typically recovered from zinc smelting residues using a combination of acid leaching and solvent extraction. As with gallium, this adds a layer of cost and complexity to the refining process, often making the economic feasibility dependent on favorable market conditions and the presence of sufficient germanium concentrations in the feedstock. Environmental factors also need to be considered, as byproduct extraction generates additional chemical waste and often involves heavy use of acids and solvents. Managing these waste streams safely is crucial, as improper handling can lead to significant environmental damage, including contamination of water resources.

The challenge in refining critical minerals lies in the fact that each mineral, along with its specific ore type, requires a tailored processing approach. These processes vary in terms of cost, efficiency, and environmental impact, adding complexity to midstream operations. REEs, for instance, require intricate chemical processes, such as solvent extraction, to isolate individual elements. This process is particularly complex because REEs often occur together in nature and must be separated one by one. Therefore, businesses that maintain continuous, optimized operations—without frequent stops or interruptions—gain significant efficiency advantages.

Building a sustainable and competitive midstream processing industry is especially challenging due to the unique characteristics of each mineral. Companies must not only advance technologies that make processing more efficient but also address the environmental consequences of these methods. Innovations like DLE for lithium and closed-loop systems that recycle chemicals and water show promise, but scaling these technologies to industrial levels remains a work in progress.

Additionally, the United States faces shortages of skilled labor and advanced equipment, as much of the processing infrastructure has moved overseas. The need for specialized knowledge in metallurgy, chemistry, and engineering further complicates efforts to scale midstream processing domestically. Developing these capabilities will require significant investment, innovation, and a strategic focus on workforce development and infrastructure expansion.

The Workforce Challenge

One of the primary challenges to expanding midstream processing capabilities in the United States is the shortage of skilled labor. Successfully processing critical minerals requires specialized expertise in fields such as metallurgy, chemistry, and engineering. Additionally, large-scale industrial operations demand experienced workers who can manage the complex processes involved in transforming raw materials into refined, usable products. Currently, the U.S. workforce is underprepared to meet these demands.

A major factor contributing to this shortage is the decline in the number of mining and metallurgical engineers. According to a 2023 McKinsey report, the number of mining engineering graduates in the United States has dropped by 39 percent since 2016.[415] Similar trends can be seen across other technical disciplines. At the same time, rapidly growing industries like battery manufacturing, semiconductors, and clean energy are attracting skilled workers with higher wages and better career opportunities. By comparison, critical minerals processing is a niche sector with fewer established players and a steeper learning curve, making it less appealing to prospective workers.

This gap in expertise is not limited to technical positions; it also affects management and operational roles. In countries like China, government-backed investments and years of experience have fostered a self-sustaining industry, complete with a robust pipeline of skilled talent. The United States, by contrast, often must rely on foreign experts or make significant investments in retraining domestic workers. This process is time consuming and costly, creating another obstacle to scaling midstream operations.

Without a coordinated national strategy to develop the necessary workforce, the United States will continue to face difficulties in expanding its midstream processing capabilities. Addressing this challenge requires investment in education and training programs that focus specifically on the skills needed for critical minerals processing. These programs must also work to increase interest in the field among new graduates and career switchers. Without a concerted effort to build a skilled workforce, the United States will face ongoing barriers to creating a strong, competitive midstream sector capable of supporting its broader industrial and technological goals.

Navigating Feedstock and Market Immaturity

The availability of feedstock for processing is a critical challenge for the U.S. midstream sector. Mining delays due to local permitting or geopolitical instability abroad in key mining regions can disrupt the steady supply of raw materials necessary to keep processing facilities operational. While some critical minerals are sourced from stable partners like Australia and Canada, others come from regions with more complex geopolitical dynamics. For instance, most of the global supply of cobalt is mined in the Democratic Republic of the Congo, a country known for political instability and humanitarian concerns. Meanwhile, the production of REEs is dominated by China, which raises additional concerns about supply chain reliability, given the geopolitical tensions between the United States and China.

A further challenge is the disparity between exchange-traded and non-exchange-traded minerals. Exchange-traded commodities, such as copper, nickel, and lithium, benefit from well-established markets with price transparency, liquidity, and hedging options. These features attract investors and create smoother market operations by allowing companies to manage risks more effectively. In contrast, non-exchange-traded minerals, like REEs and gallium, present far more opaque pricing, limited buyers, and unpredictable supply chains. This lack of transparency makes it difficult for businesses to forecast costs and plan investments, which further discourages the development of processing facilities for these minerals.

The absence of transparent and reliable markets makes it challenging for U.S. producers to attract private capital. Investors are hesitant to engage in industries where pricing and supply are unpredictable, leading to financing gaps that hinder the expansion of midstream infrastructure. This decentralized nature of the U.S. critical minerals market stands in sharp contrast to China's vertically integrated model, where state-backed investments ensure a more stable supply chain from mining through refining. In China, government support for mining and processing creates a cohesive market ecosystem that offers stability to investors and businesses alike.

For the United States to build a viable midstream sector, policy support will be essential in improving the transparency and maturity of domestic markets for critical minerals. This could include measures such as those described below to increase price transparency, encourage the establishment of standardized markets for non-exchange-traded minerals, and provide incentives that make investment in these areas more attractive. Without these steps, U.S. efforts to establish a robust midstream processing capability will continue to face significant barriers due to market uncertainties and the unpredictability of feedstock availability.

POLICY LEVERS

How the United States Can Close the Gap

To tackle the challenges of developing midstream processing capabilities for critical minerals in the United States, a multifaceted strategy is required—one that integrates government policy with private sector innovation. Several key policy actions can help accelerate progress in this area.

1. **Support workforce development.** Establishing a National Critical Minerals Workforce Initiative will address workforce challenges in domestic projects by leveraging federal programs and tax incentives. The Department of Labor should prioritize critical minerals in the Workforce Innovation and Opportunity Act (WIOA) to support education in metallurgy, chemistry, and mining engineering. Community colleges can partner with WIOA boards to create certifications and degrees, while regional training centers can integrate into the American Job Centers network, enhanced by the National Science Foundation's Advanced Technological Education program.

   The Carl D. Perkins Career and Technical Education Act can also support high school programs in critical minerals, building a talent pipeline. Funding for science, technology, engineering, and mathematics (STEM) clubs and science fairs can promote interest, and collaboration with the Department of Energy's Minority Educational Institution Student Partnership Program can encourage underrepresented groups.

   Expanding scholarships and loan forgiveness programs will attract talent. The Teacher Education Assistance for College and Higher Education (TEACH) Grant Program can offer scholarships for mining and chemical engineering, while a loan forgiveness program can support graduates in the critical minerals sector, with coordination from the Department of Defense to create a fund under the department's Science, Mathematics, and Research for Transformation (SMART) program.

2. **Establish the U.S. Critical Minerals Reserve.** The United States should establish a U.S. Critical Minerals Reserve to ensure a stable and secure supply of essential raw materials needed for midstream processing, thereby strengthening national security and economic resiliency. This reserve would function as a financial and strategic mechanism to mitigate supply disruptions caused by geopolitical risks, market volatility, and timing misalignments that occur due to the noncontiguous ramp of mining, processing, downstream production, and supply and demand. This approach aligns with the recommendations from the House Select Committee on the Strategic Competition Between the United States and the Chinese Communist Party, which emphasized reducing U.S. dependence on foreign-controlled supply chains for critical minerals, especially those dominated by China.[416]

   The mechanism by which the reserve would work is straightforward. By lending capital to authorized market makers (AMMs), who will then procure specified critical minerals from allowed jurisdictions, the reserve would support domestic and allied supply chains, stabilize mineral prices, and mitigate risks associated with foreign-controlled resources. The AMMs, backed by private capital to achieve profit, would seek to reduce production costs through innovation while supporting downstream efforts to grow demand. The AMMs would take price risk, backed by federal loans, shifting that risk away from U.S. critical minerals producers. Consequently, the industry would gain resilience.

3. **Update permitting to reduce delays and preserve safeguards.** To facilitate the development of critical minerals projects, it is important for the U.S. federal government to prioritize efforts to update the permitting process for mining and midstream operations. In order to minimize bureaucratic delays while preserving essential environmental safeguards, enhanced coordination between different federal, state, and local agencies will be required. In doing so, a more efficient pathway for obtaining permits can be established, which is crucial for ensuring a stable and competitive supply of critical minerals.

   A central component of this approach involves building on the existing Federal Permitting Improvement Steering Council (FPISC) to create a centralized "one-stop shop" specifically for critical minerals projects. Presently, navigating the permitting landscape requires companies to interact with multiple regulatory bodies, often leading to redundancies and significant delays. By leveraging the framework of FPISC, a more cohesive system can be put in place—one that aligns regulatory agencies at all levels and reduces inefficiencies. In 2021, mining was added to FPISC's list of covered projects under the Fixing America's Surface Transportation Act (FAST-41).[417] Expanding the FAST-41 program to include critical minerals projects explicitly would be instrumental in ensuring dedicated personnel are available to manage these projects, providing necessary technical assistance and promoting transparency in the permitting process.

Another key strategy is to increase the use of programmatic environmental reviews and categorical exclusions to expedite the permitting process for critical minerals projects at the federal level. Programmatic environmental impact statements can be developed to evaluate the impacts of broad categories of activities associated with critical minerals in advance rather than requiring repetitive, project-by-project reviews. By covering common types of critical minerals activities—such as mineral exploration, drilling, processing, and material manufacturing—these programmatic reviews can establish baseline analyses. This proactive approach streamlines approvals by allowing subsequent individual assessments to focus on specific site conditions and deviations rather than reanalyzing general impacts that have already been addressed.

An interagency task force led by the Department of the Interior and including the Environmental Protection Agency, Bureau of Land Management (BLM), and the U.S. Forest Service could provide the structure needed to improve the permitting process. This task force would prioritize developing clear timelines for permit applications, allocating resources to support permit reviews, and offering guidance to companies seeking permits. Additionally, increased funding for the BLM and U.S. Geological Survey will be important so that these agencies can hire specialized personnel devoted exclusively to critical minerals projects, with the goal of reducing approval times without compromising the integrity of environmental assessments.

Amending the America Creating Opportunities for Manufacturing, Pre-Eminence in Technology, and Economic Strength (COMPETES) Act to mandate the expedited consideration of critical minerals projects, particularly those with implications for midstream processing, could serve as a platform for enacting legislation related to critical minerals permitting. By adopting a model similar to the Renewable Energy Coordination Offices within the BLM, a specialized expedited permitting pathway could be established for critical minerals, ensuring that these projects receive appropriate priority.

By creating a streamlined pathway and prioritizing projects of strategic importance, the United States can foster a resilient and competitive critical minerals sector that is better positioned to meet current and future demand.

CONCLUSION

A Path Forward for U.S. Midstream Leadership

To close the midstream gap, the United States must pursue a coordinated strategy that combines policy support, private investment, technological innovation, and workforce development. As the global race for critical minerals intensifies, the United States cannot afford to fall behind in building a robust processing and refining capability. Without it, the country risks not only lagging in technological advancements but also compromising its energy security and economic independence.

The path forward to establishing a competitive critical minerals midstream in the United States will be challenging. Labor shortages, the need for market maturation, and reliance on foreign suppliers all present significant hurdles. However, the opportunities for U.S. leadership are equally significant. By fostering innovation, investing in skilled talent, and building resilient, transparent supply chains, the United States can strengthen its position as a global leader in critical minerals. This effort is essential for securing a stable foundation for the future of technology, and by acting decisively now, the United States can ensure long-term competitiveness in the global market.

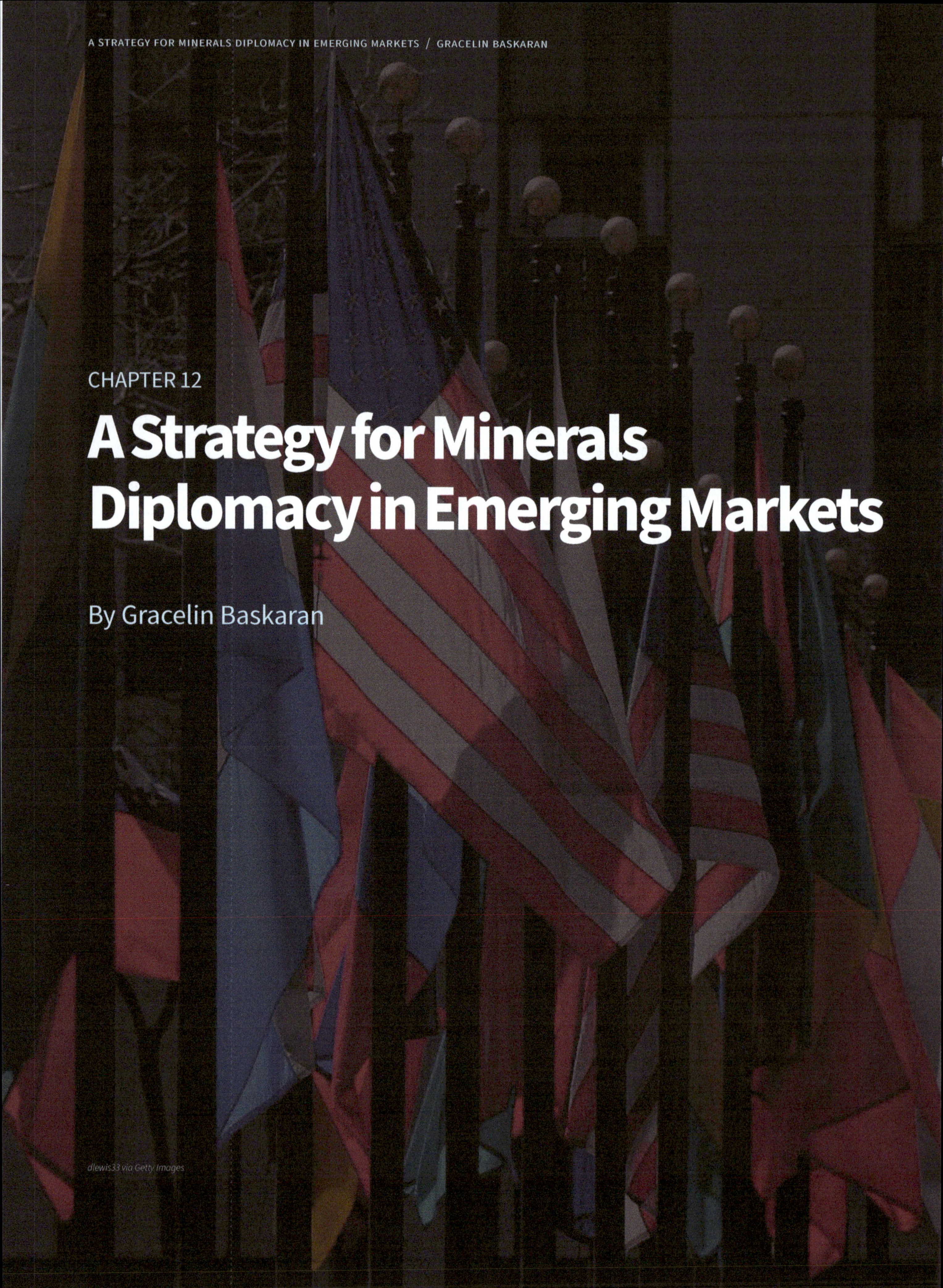

CHAPTER 12

A Strategy for Minerals Diplomacy in Emerging Markets

By Gracelin Baskaran

dlewis33 via Getty Images

The United States cannot, and should not, remain on the sidelines when it comes to minerals. But its entry into the international field will have to come with a model that competes on economics and practices.

Over the past 30 years, China has emerged as a critical player in the mineral supply chains crucial for national and energy security. Although it produces only 10 percent of the world's lithium, cobalt, nickel, and copper, China imports sufficient quantities to process 65 to 90 percent of the global supply of these metals.[418] This dominance is the result of years of industrial planning and foreign policy initiatives from Beijing. For the United States, China's dominance poses a strategic challenge, especially given rising geopolitical tensions and Chinese export restrictions on critical commodities, including gallium and germanium, which are vital for semiconductors; graphite, which is key for electric vehicles; and antimony, which is used in many defense technologies.

Because the United States has limited domestic reserves—including less than 1 percent of the world's reserves of commodities such as cobalt, nickel, and graphite—it must develop a strategy to reduce its dependence and enhance its mineral supply security. The good news is that many resource-rich countries want to work with the United States and its partners to develop mineral resources if they can bring tangible benefits to host jurisdictions. Resource-rich countries want to diversify partners because they have realized that relying on a single country for investment and offtake is a risk. Additionally, Western companies are generally more responsible, environmentally conscious, and attentive to human rights and labor conditions, which can reduce the friction between communities, workers, firms, and the government. The bad news is that the U.S. government cannot command the efforts of mining companies, which are private companies accountable to shareholders. Instead, the United States must create an environment that will enable the private sector to compete with state-owned enterprises and offer more benefits to resource-rich countries. The result should be a new model of mining that is mutually beneficial for companies, resource holders, and U.S. consumers.

Both the Trump and Biden administrations set out to improve supply through better exploration, production, recycling, and reprocessing of critical minerals. However, the United States' efforts to secure its supply chain have been reactive and unsystematic, especially in international exploration and mining. U.S. officials often learn about projects after they have been announced, missing out on the chance to coordinate investments with Western partners. In particular, the United States will need to engage in Africa, Latin America, and other resource-rich regions in the Global South.

Many of the world's most resource-rich countries are among the most difficult in which to do business. U.S. government efforts can enable investment by helping reduce risks for projects and operators. The solutions, however, are not one-size-fits-all. A framework for prioritizing countries is a key starting point—and then identifying how tools can be deployed in these jurisdictions can follow. Approaches can include government-to-government cooperation agreements, technical assistance on legal and regulatory reforms, public-private partnerships, and concessional financing.

It will also be important to overcome the historical legacy of mining, wherein communities and workers bore many of the consequences of resource extraction, including environmental damage, human displacement, and poor economic growth.

The United States cannot, and should not, remain on the sidelines when it comes to minerals. But its entry into the field will have to come with a model that competes on economics and practices. With a more systemic approach, the United States can enable its diplomats, policymakers, and private sector actors to source the minerals needed to process and manufacture goods that are required for the twenty-first century.

The U.S. government has limited resources—and developing a targeted strategy for engagement will require identifying priority jurisdictions and developing appropriate tools to advance U.S. interests. Given the ever-changing landscape of electoral outcomes, policy changes, and geological discoveries, this chapter establishes criteria for identifying priority jurisdictions—drawn from ample global experience—and makes policy recommendations on how the United States can use financial and nonfinancial tools to advance its goals.

A FRAMEWORK FOR IDENTIFYING PRIORITY JURISDICTIONS FOR GOVERNMENT COOPERATION

To date, the United States has primarily engaged with emerging markets through aid and national security initiatives. In 2007, the United States established the Africa Command to address security challenges, with the U.S. military spending nearly $2 billion annually on its operations in Africa.[419] On the aid front, in 2023, U.S. government aid to emerging economies totaled $223.7 billion, of which $53.5 billion was allocated to Africa.[420] Latin America received considerably less—approximately $2.5 billion was requested by the Biden administration for Latin America and the Caribbean in the same year, to be disbursed through the State Department and the U.S. Agency for International Development (USAID).[421]

Nonetheless, the United States has done little to engage with these countries from a commercial perspective, particularly in the mining sector. Recent efforts include the Lobito Corridor under the Partnership for Global Infrastructure and Investment, a $150 million loan for a graphite project in Mozambique, and backing through the Minerals Security Partnership for the Serra Verde rare earth elements project in Brazil.[422] However, these investments have not followed a cohesive strategy and generally occur in isolation.

There are five broad considerations that determine a high-potential mining jurisdiction: (1) geology—mineral reserves and quality; (2) policy stability, transparency, and rule of law; (3) quality of infrastructure; (4) the ability to secure and maintain a social license to operate; and (5) historical trade relations and market access. This section draws on examples from across the world to unpack these considerations. However, none of these factors can be taken in isolation; prioritization should be based on their aggregate impact.

Quantity and Quality of Mineral Reserves

Any strategy must align priority minerals with available resources, taking into consideration recent discoveries, ore grade, and the size of identified deposits. Indonesia serves as an excellent example of a jurisdiction prioritized by investors and policymakers despite its challenging policy environment, driven by rising resource nationalism. Not only does Indonesia have a fifth of the world's nickel, but it also has some of the highest-quality and most profitable reserves, as is evident through metrics like recovery rate and mill-head grade.[423] (Recovery rate refers to the proportion of nickel successfully extracted from raw ore, while the mill-head grade indicates the average nickel concentration in mined ore processed through a mill.) Among the world's five largest nickel mines—Kola Division in Russia, Jinchuan in China, Sudbury Operations in Canada, Sorowako in Indonesia, and Polar Division in Russia—Sorowako stands out with the highest recovery rate (88 percent, compared to 25.4–85 percent for the other four mines) and the second-highest mill-head grade (1.68 percent), surpassed only by Russia's Kola operation (2.3 percent).[424] Indonesia's dominance in nickel production played a significant role in U.S. president Joe Biden and Indonesia's then-president Joko Widodo's decision to elevate their countries' bilateral relationship to a Comprehensive Strategic Partnership in 2023, despite Indonesia's poor environmental track record in nickel production.[425]

Policy Stability, Transparency, and Rule of Law

Mining investments are long term, so confidence in the stability of the government and its policies; institutional capabilities to enforce mineral legislation, regulations, and standards; and transparency are all important.[426]

Despite possessing some of the world's richest copper deposits—boasting ore grades exceeding 3 percent, substantially higher than the global average of 0.6–0.8

percent—and holding half the world's cobalt reserves, the Democratic Republic of the Congo (DRC) struggles to attract Western mining companies due to political instability, corruption, and weak enforcement of laws.[427] At present, Glencore and Ivanhoe are among the only Westerns firms operating in the DRC. Both First Quantum Minerals, a Canadian company, and Freeport-McMoRan, a U.S. company, have exited since 2010.[428] First Quantum Minerals' departure from the DRC resulted from a conflict with the government over the expropriation of one the company's major assets.[429] During a visit to the DRC in August 2024, both industry and government officials agreed that corruption is a significant deterrent to attracting Western investment. Without a central tax authority, any government entity can impose a tax and freeze a company's assets until it is paid. So, despite having some of the best copper deposits in the world—both in size and quality—and half the world's cobalt, investors remain hesitant.

Policy stability is a key consideration. As demand for minerals rapidly rises, an increasing number of governments are adopting resource nationalism policies to secure a larger share of benefits from their natural resources. These policies may include higher taxes and royalties, nationalization or state ownership, export control (such as nonautomatic licensing, quotas, tariffs, or bans), and local ownership requirements. In recent years, countries have leveraged bans on raw resource exports to mandate local processing and value addition. However, such policies often elicit mixed reactions from investors. New investors may hesitate to enter, while existing investors must decide whether to comply or divest. Countries that adopt resource nationalism policies on short notice risk signaling policy volatility, which can deter investment. Greenfield investment is drying up in jurisdictions with resource nationalism.

Quality of Infrastructure

Mining and processing require critical infrastructure, including energy, water, ports, rail, and roads. Mining is one of the most energy-intensive industries in the world, accounting for 10 percent of global energy consumption.[430] Estimates suggest that the mining industry's energy consumption could grow by a factor of two to eight by 2060.[431]

Affordable and stable base load power can make mining and processing investments significantly more commercially attractive. Saudi Arabia has made substantial investments in renewable energy and boasts some of the lowest energy costs in the world. Its Al Ghat wind project set a record for the lowest electricity cost from wind power—just 1.56558 cents per kilowatt hour (kWh) levelized cost of energy (LCOE). Saudi Arabia's Wa'ad Alshamal project has the second-lowest cost, at 1.70187 cents/kWh LCOE.[432] Additionally, Saudia Arabia has the lowest LCOE for solar photovoltaics.[433] These developments have positioned Saudi Arabia to become one of the world's largest mineral processing hubs. By 2030, the kingdom aims to be one of the seven biggest mineral processors in the world.

Countries with energy shortages are at a disadvantage when it comes to attracting mineral extraction and processing investments. Some of the most mineral-rich countries are among the most energy-poor—and they are overwhelmingly in Africa. In 2022, only 21.5 percent of the population in the DRC had access to energy, compared to 33.2 percent in Mozambique, 36.1 percent in Madagascar, and 45.8 percent in Tanzania.[434] While some companies have opted to build their own energy generation and transmission infrastructure, this adds a significant cost to doing business. In South America, the global mining company Anglo American has made significant energy investments and operates entirely on renewable energy.[435]

Transportation infrastructure is vital for global mineral supply chains. Commodities must move within countries and across borders and oceans. Among the world's 10 largest ports by volume of goods moved, 7 are in China. Of the 50 largest ports, 2 are in Latin America (Santos, Brazil; and Colon, Panama), and one is in Africa (Tanger Med, Morocco). These ports serve as critical hubs facilitating the global trade of minerals.[436]

Rail is the most efficient way to transport minerals to ports, as the bulk nature of commodities can be damaging to roads. Transporting bulk commodities by rail also generates 75 percent fewer carbon emissions compared to road journeys.[437] Africa's landmass is larger than the combined areas of the continental United States, China, Europe, and India—yet its 82,000 kilometers of rail is only slightly more than the combined rail networks of France and Germany.[438] Much of Africa's rail infrastructure has been poorly maintained, with nearly a fifth entirely nonoperational. As a result, most commodities are transported by road. However, roads also present significant challenges. Africa has one of the lowest road densities in the world, with just 27 kilometers of road per 10,000 people, and less than a third of its roads are paved.[439] This lack of infrastructure underscores the rationale behind the Partnership for Global Infrastructure Investment's Lobito Corridor, which aims to connect the DRC, Zambia, and the rest of Angola to Angola's Port of Lobito, thereby reducing transportation costs for mineral exports.[440]

Securing Social Licenses to Operate

Social license to operate is a concept used to describe the informal and ongoing approval of a mining company's operations by the surrounding community and stakeholders. Unlike legal permits issued by governments, social licenses are granted by communities and based on trust, legitimacy, and consent. Social license to operate is one of the biggest risks facing mining companies. For three consecutive years—2019, 2020, and 2021—it topped the list of the 10 biggest risks cited by the mining industry in Ernst & Young's annual survey.[441] Half of the mining executives surveyed identified it as the biggest risk.[442] Looking ahead, executives still rank it among the top five risks facing the industry in 2025.[443]

Peru is a prime example of the impact of social license to operate on investment. It is one of the only Latin American countries to have seen a decline in investment in recent years. Over the last five years, from 2019 to 2024, exploration investment increased by 21.6 percent in Chile, 26.9 percent in Mexico, 51.7 percent in Brazil, and 105.4 percent Argentina. The only major mining jurisdiction that saw a reduction in exploration investment was Peru, which declined by 12.6 percent.[444] During this time, there has been a substantial increase in social unrest related to mining in Peru. In August 2023 alone, there were 71 mining-related protests.[445] Fitch Ratings, a major credit rating agency, cited mining-related social unrest, protests, and blockades as a significant challenge.[446]

Panama is increasingly viewed as "uninvestable" after social unrest and subsequent government action led to the closure of its biggest mine. First Quantum Minerals' $10 billion Cobre Panama copper mine was one of the world's largest copper-producing mines, accounting for 1.5 percent of global copper supply. Since beginning production in 2019, it had accounted for 75 percent of Panama's exports and 5 percent of its gross domestic product. However, a storm was brewing. In addition to high unemployment and corruption there was also growing resentment over the mine's environmental impact—particularly river and soil contamination.[447] There was also sentiment that the mine should be in the hands of the Panamanian people rather than foreign companies. Illegal blockades of the international port and mine access road forced the mine to stop production in November 2023. These were the country's worst protests in decades. At the end of November 2023, the Supreme Court of Justice ruled that First Quantum Minerals' mining concession contract was unconstitutional.[448] It is hard to see an outcome in which mining companies would want to undertake new projects in Panama.

Historical Trade Relations and Market Access

Market access—or a company's ability to sell goods and services across borders—is a critical determinant of a company's investment decisions. Market access involves the conditions, barriers, and regulations that impact trade, such as tariffs, quotas, technical standards, and trade agreements. A strong history of bilateral trade—built on stable market access—can make mining investments more commercially attractive to investors. Canada, China, the European Union, Japan, Mexico, and the United Kingdom are among the United States' biggest trading partners. With the exception of China, which has imposed a barrage of export restrictions, there is little concern that future investments in these countries are at risk. Historical trade ties are a strong signal.

Namibia is a prime example of a country with which the United States has strong trade ties—and significant room to increase mineral trade. Namibia is a key beneficiary of the African Growth and Opportunity Act, a U.S. unilateral trade preference program signed into law by President Bill Clinton in 2000 and renewed in 2015. Importantly, Namibia has key critical minerals vital to safeguarding U.S. security interests. Its Lofdal heavy rare earths operation produces 2,000 tons per year of rare earth oxides and has rich deposits of two of the most valuable heavy rare earth metals, which are used in defense systems, lasers, electronics, and renewable energy technologies. Namibia also has the third-largest uranium reserves in the world—a vital input for nuclear power—and has large copper, cobalt, lithium, zinc, and fluorspar reserves.[449]

LEVERAGING THE FRAMEWORK FOR MINERALS DIPLOMACY

When determining which countries the United States should prioritize, none of the considerations discussed above should be taken in isolation. If geology alone were enough, policymakers and investors would focus on mining nickel in Australia, whose total reserves match those of Indonesia. Unfortunately, higher production costs have rendered Australian nickel uncompetitive.[450] Over the last two years, BHP closed its Nickel West operations and West Musgrave project in Australia, and Glencore shut down its Koniambo Nickel SAS operation in New Caledonia due to unprofitability.[451]

If historical trade ties were enough, the United States would have much stronger minerals trade with Latin America. The United States has more free trade agreements with Latin American countries than with any other region in the world. These include agreements with Chile, Mexico, and Peru. Yet, in 2021, Chile exported $28.2 billion worth of copper to China, compared to just $6.5 billion to the United States. Similarly, Mexico, one of the United States' biggest trade partners, sent 94 percent of its copper exports to China, compared to just 1 percent to the United States.[452]

If resource nationalism were the sole determining factor, Vale Base Metals would cease mining nickel in Indonesia, and Freeport-McMoRan would not still have its copper operations there.

If social license to operate were the determining factor, First Quantum Minerals would not be focusing its efforts on restoring operations at its Cobre Panama mine.[453]

If infrastructure was the primary determinant, investors would not be flocking to Zambia. The country is experiencing a 500 megawatt (MW) energy deficit due to a drought that is undermining hydroelectric power generation and causing rolling blackouts. The country also has significantly higher transportation

Figure 1: Chile's Copper Exports, 2001–2021

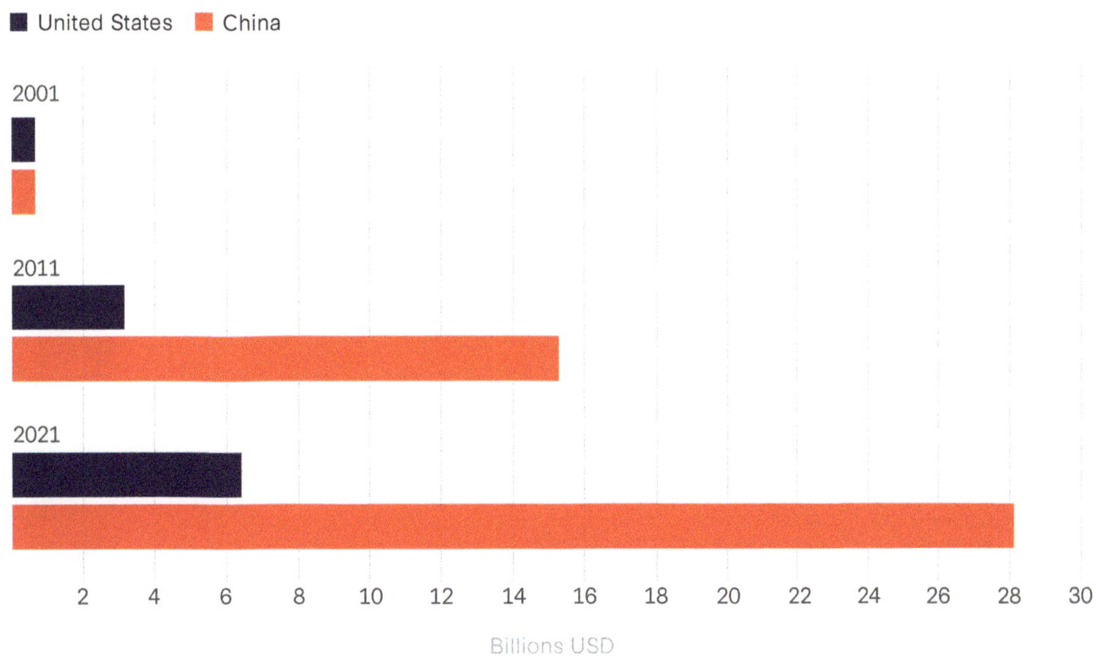

Source: CSIS analysis.

costs because it is landlocked. Transportation costs add 40 percent to production expenses because accessing ports such as Beira, Dar es Salaam, Durban, Lobito, and Walvis Bay is difficult and time consuming.[454] Nonetheless, First Quantum Minerals announced a $1.3 billion investment to expand copper production; Anglo American has returned to Zambia after two decades; Rio Tinto is pursuing an exploration campaign; KoBold, a U.S. firm, is building Zambia's biggest copper mine; IHC has acquired a major asset; and Barrick Gold has scaled up exploration with plans to create a super-pit at its existing mine, potentially extending its life by 60 years.[455]

It is the aggregation of these considerations that determines what constitutes a high-potential mining jurisdiction.

Table 1 evaluates the performance of selected countries, chosen because they are home to multiple priority resources. The table applies the considerations framework to these countries.

Canada, Australia, and Saudi Arabia are high-income countries with which the United States should pursue collaboration, including providing financing and leveraging subsidies and tax credits, as outlined in the table above. When it comes to emerging markets, Namibia should be a priority jurisdiction. With some infrastructure support, Tanzania and Zambia's attractiveness could increase substantially, making them important sources of nickel, graphite, and rare earth elements. Argentina, Brazil, and Chile would benefit from building capacity for firms to strengthen their social license to operate through more active benefit sharing with communities. It is no surprise that there is an aversion among Western investors to enter the DRC and even South Africa, or that Peru is one of the only Latin American countries to have experienced a decline in investment over the last five years. Addressing these factors can improve the attractiveness of these jurisdictions to mining companies.

Table 1: Performance Across Key Investment Indicators in Resource-Rich Mining Jurisdictions

	Criticality of resources*	Policy climate, stability, and rule of law	Quality of infrastructure	Social license to operate	Historical trade relations[456]
Argentina	Lithium Copper				
Brazil	Rare earths Nickel Graphite Manganese Titanium Niobium Aluminum Copper				
Chile	Copper Lithium				FTA**
Mexico	Copper Bismuth Graphite Manganese Zinc				FTA
Peru	Copper Bismuth Zinc				FTA

	Criticality of resources*	Policy climate, stability, and rule of law	Quality of infrastructure	Social license to operate	Historical trade relations
Tanzania	Copper Cobalt Nickel Graphite Rare earths Uranium				AGOA***
South Africa	Platinum Palladium Manganese Chromium Copper Manganese Uranium Beryllium Titanium Vanadium Nickel Rare earths Fluorspar Zinc				AGOA (at risk of losing based on recently introduced legislation)
Namibia	Rare earths Lithium Copper Cobalt Magnesium Zinc Uranium				AGOA
Democratic Republic of the Congo	Copper Cobalt Tantalum Lithium Tungsten				
Zambia	Copper Cobalt Nickel				AGOA
Indonesia	Nickel Copper				
Saudi Arabia	Copper Aluminum				

	Criticality of resources*	Policy climate, stability, and rule of law	Quality of infrastructure	Social license to operate	Historical trade relations
Canada	Nickel Cobalt Aluminum Fluorspar Indium Niobium Palladium Platinum Tellurium Titanium Uranium				FTA
Australia	Nickel Antimony Zinc Uranium Cobalt Copper Lithium Tungsten Vanadium Aluminum Magnesium Titanium				FTA
United States	Copper Platinum Zinc Titanium Magnesium Beryllium Lithium Nickel				N/A

Source: Author's elaborations based on interviews with government and industry, mine visits, S&P CapitalIQ, and data from the Global Infrastructure Hub, Fraser Index, Rule of Law Index, ACLED, World Economic Forum, and the World Bank.

* The criticality of resources is determined by the specific resources a country possesses rather than the in-situ value of all resources (e.g., including gold). Countries with resources that the United States has limited reserves of and that are vital for U.S. security interests are ranked higher.

** Free trade agreement

*** African Growth and Opportunity Act

RECOMMENDATIONS FOR BUILDING MINERALS DIPLOMACY

The United States has a range of tools it can better deploy to strengthen minerals diplomacy. While discussions often center on concessional financing, the United States' arsenal is significantly larger. If used strategically, these tools can enhance U.S. supply security and redesign supply chains, enabling countries that have historically exported resources to China to shift their exports to the United States and other Western nations. These instruments can be broadly categorized into four areas: financing, geological mapping, infrastructure, and market incentives. This section offers recommendations for each of these four areas.

1. **Reform the U.S. International Development Finance Corporation.** The U.S. International Development Finance Corporation (DFC) was not originally established to finance minerals security needs. However, it has become the primary vehicle for funding such projects overseas.

Two amendments could significantly enhance the DFC's capacity to support minerals security interests. First, the White House's Office of Management and Budget should amend its rules to allow the DFC to make equity investments in mining projects. Equity investments send a strong signal to both companies and countries, as government equity can help mobilize additional private capital and foster government-to-government cooperation. Second, revising or introducing legislation to enable the DFC to finance mining projects in high-income, resource-rich countries is essential. Under current rules, the DFC cannot fund projects in countries such as Canada and Chile. Congress could address this issue similarly to its approach with the European Energy Security and Diversification Act of 2019. That legislation authorized the DFC to finance energy projects in high-income European countries, including a $500 million loan guarantee for a liquified natural gas project in Poland aimed at reducing reliance on Russia.[457]

2. **Support infrastructure development in priority jurisdictions.** One way China has advanced its minerals security goals is through infrastructure development. This approach benefits host jurisdictions while ensuring that mining operations have access to essential energy, water, and transportation infrastructure. Such investments are vital for sustaining the mining ecosystem. During the first 10 years of China's Belt and Road Initiative (BRI), it invested $1 trillion, including $634 billion in construction contracts and $419 billion in nonfinancial investments.[458] In 2023 alone, China allocated $7.9 billion to energy investments. That same year, it made BRI investments in 61 countries in 2023, with Africa overtaking the Middle East as the largest recipient.[459] These investments have allowed China to secure raw materials from domestic processing. For example, in 2018, Chile became one of the first Latin American countries to join the BRI, and today, nearly one-third of China's raw copper comes from Chile.[460]

   The United States has several mechanisms for financing infrastructure abroad, including the DFC and Power Africa. Power Africa, launched by President Barack Obama in 2013, was an innovative, government-led partnership aimed at increasing energy access in Africa. Initially, it set out to add 10,000 MW of energy capacity and 20 million connections by 2030. In 2014, Power Africa tripled its goals.[461] However, over time, budgetary appropriations have significantly declined, hampering its ability to operate at the required scale.

   To deploy soft power effectively, strengthen minerals diplomacy, and reduce barriers to entry for Western mining companies, mechanisms like the DFC and Power Africa will require substantially larger budget allocations.

3. **Leverage the U.S. Geological Survey to undertake geological mapping and de-risk exploration in priority jurisdictions.** Many developing countries remain either unmapped or reliant on outdated geological surveys. For example, 45 percent of Zambia's land is still geologically unmapped, and the remaining 55 percent was last mapped before 1998. Geological mapping can help de-risk investments for exploration companies, which is crucial given that over 99 percent of exploration projects are unsuccessful.[462]

   Building mapping capabilities will require budgetary allocations. Additionally, embedding U.S. Geological Survey attachés at embassies to oversee mapping efforts presents an important opportunity for advancing minerals diplomacy. These attachés could collaborate with geological surveys and mining ministries in host jurisdictions, fostering stronger partnerships and supporting informed resource development strategies.

4. **Develop carrots—subsidies and tax credits—to incentivize investment in priority jurisdictions globally.** The lack of U.S. investment incentives has allowed China to gain control over many of Africa's natural resources. China's foreign direct investment in Africa grew from $75 million in 2003 to $4.2 billion in 2020, primarily in the mining sector.[463] Similarly, China has made substantial investments in Latin America, focusing on critical minerals, energy, telecommunications, and transport infrastructure. In November 2024, Chinese president Xi Jinping visited Peru, where he launched the first phrase of a $3.5 billion port project aimed at strengthening trade routes from Latin America's Pacific coast to China.[464]

Meanwhile, the value of trade between China and Africa increased from $10 billion in 2000 to $25 billion in 2021—over four times the growth seen between the United States and Africa.[465] In Latin America, Chinese trade has soared from $12 billion in 2000 to $450 billion in 2023.[466] Investment incentives and de-risking measures—such as subsidies, tax credits, and loan guarantees—will be critical for countering Chinese investment and fostering U.S. engagement in these strategic regions.

CONCLUSION

While an "America First" approach and developing domestic mining are important, continuing to lag in emerging markets presents a critical vulnerability for the United States for two reasons. First, the United States lacks sufficient quantities and quality of many commodities essential for national, economic, and energy security. Second, allowing China to continue to outpace the United States in Latin America, Africa, and Asia only strengthens China's dominance in global mineral supply chains.

However, the United States cannot pursue efforts in every resource-rich country. Prioritizing target countries for minerals diplomacy will be crucial. This chapter proposed a framework for identifying these countries based on five considerations: (1) mineral reserves and quality; (2) policy stability, transparency, and rule of law; (3) quality of infrastructure; (4) the ability to secure and maintain a social license to operate; and (5) historical trade relations and market access.

Ultimately, the United States must move beyond discussions about what it should do—action on financing instruments, infrastructure and mapping, and incentives is imperative.

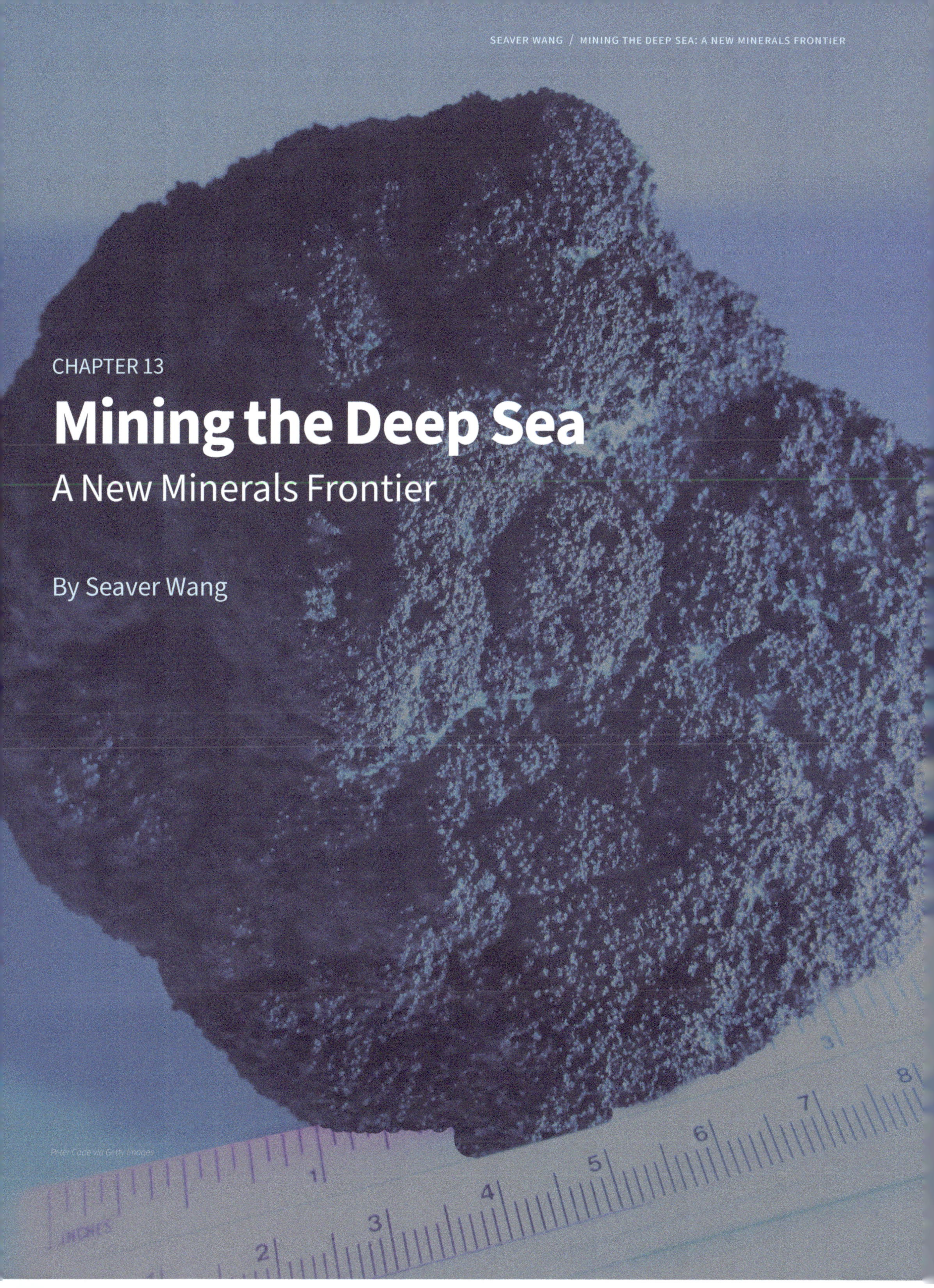

CHAPTER 13

Mining the Deep Sea

A New Minerals Frontier

By Seaver Wang

Peter Cade via Getty Images

China's multi-sector deep sea mining industry efforts, which enjoy clear national policy backing, stand in stark contrast with the United States' nonexistent seafloor minerals strategy.

In the coming years, scientific, technological, and regulatory trends appear likely to unlock the economically viable and environmentally responsible collection of metals-rich nodules from the abyssal seafloor at depths of several kilometers. Yet despite the potential of seafloor polymetallic nodules to dramatically alter existing critical minerals supply chains, the United States has neglected to develop policies that can help realize the potential of this emerging sector and deliver strategic national benefits. At present, U.S. passivity increases the likelihood that new production from seafloor metals will simply flow to Chinese metallurgical processing plants, exacerbating current patterns of critical minerals import dependence.

A robust U.S. critical minerals strategy should immediately break from this trend of inaction and take steps to leverage nodule resources as a new, diversified source of metals to support energy transition efforts. In the near term, given its nonparticipation in the UN Convention on the Law of the Sea (UNCLOS), the United States cannot directly access minerals within the international seabed area. Policymakers should therefore take a multi-pronged approach, aligning diplomatic efforts, existing Inflation Reduction Act (IRA) incentives, and policy support for downstream industries to attract nodules extracted from overseas to domestic mineral-processing and battery-manufacturing plants. Eventually, the United States should become a party to UNCLOS, gaining equal access to minerals from the international seabed and helping influence and strengthen international governance of the high seas. Public support for research on the seabed, deep sea technologies, and environmental best practices can also better position the country to compete in this emerging sector. Meanwhile, improved federal mapping of offshore mineral resources in territorial and extended continental shelf waters may identify promising mineral deposits for development under U.S. sovereign authority.

THE NEW MINERAL FRONTIER

With each passing year, viable commercial mineral harvesting from the seafloor becomes increasingly realistic. Historical barriers to deep sea mining are diminishing in the face of better resource mapping, improving robotics, and falling cost propositions relative to land-based resources. Spurred by growing demand for minerals for clean energy, regulatory developments for resource collection on the international seafloor are proceeding slowly but steadily. Countries are also increasingly pursuing seafloor mineral resources in their exclusive economic zones (EEZs).

In 2024, the Norwegian government moved to allow exploratory resource-mapping and equipment-testing activities on its offshore continental shelf.[467] Japan has tested the viability of several different kinds of seafloor mineral deposits in its EEZ in past years, most recently discovering a significant concentration of metals-rich nodules in the country's easternmost sovereign waters in July 2023.[468] In August 2024, India opened up its EEZ for exploration contracts.[469] In the South Pacific, the Cook Islands expects to make key regulatory decisions on seafloor nodule collection within years.[470] Meanwhile, in December 2023, the United States took a key first step toward securing seafloor resources by declaring boundaries for the U.S. extended continental shelf (ECS).[471]

Under UNCLOS, seabed minerals and extractive activities beyond the boundaries of countries' sovereign rights fall under the regulatory authority of the International Seabed Authority (ISA), an independent international organization. However, opposition from lawmakers hostile to international seafloor governance has blocked the U.S. Senate from ratifying UNCLOS for over 30 years, leaving the United States one of the few countries not party to UNCLOS and the ISA. Without full membership, it cannot sponsor mineral-related activities on the international seabed. At the same time, the federal government has devoted little policy attention to pursuing hardrock minerals in sovereign waters. Thus, despite seafloor metals' value as an alternative to import dependence on critical minerals from China, the United States still lacks any meaningful policy strategy for ocean minerals. Without decisive federal action, seabed minerals may simply become the newest entry on a long list of strategic commodities firmly under Chinese state and industry control.

Three different types of mineral-rich seafloor deposits have historically attracted the most attention for excavation: polymetallic nodules (also called manganese nodules), ferromanganese crusts (also called cobalt crusts), and polymetallic sulfide deposits (also called seafloor massive sulfides). No commercial exploitation of these seafloor minerals is taking place today, although exploration efforts (i.e., resource mapping and equipment testing) have increased in recent years. Some polymetallic nodule and ferromanganese crust resources likely lie in seafloor areas under U.S. jurisdiction, particularly in the Pacific, but these and other deposits in U.S. EEZ waters remain incompletely characterized.[472] Globally, seabed mineral exploration to date has mostly targeted the international seafloor, particularly rich nodule fields in the central and western Pacific.

Ferromanganese crusts and polymetallic sulfide deposits are tightly embedded into seafloor rock and sediment, making extraction difficult, environmentally challenging, and reliant on excavation techniques akin to those used for surface mining on land.[473] Exploratory testing of systems to mine them remains relatively immature and may prove difficult to balance cost effectively with good environmental management. Because environmentally responsible and economically viable harvesting of ferromanganese crust and polymetallic sulfide resources is likely beyond reach for the foreseeable future, focusing policy efforts on such deposits seems relatively unproductive.

By contrast, polymetallic nodules are potato-sized nuggets of metallic minerals that sit clustered atop sediment-covered, plains-like regions of the abyssal seafloor, typically at depths of 3.5–6 kilometers (2.2–3.7 miles).[474] These nodules possess rich manganese, nickel, copper, and cobalt content.[475] Collection requires little more than a means of lifting them to the ocean's surface. Most collector system concepts involve using robotic underwater vehicles that separate nodules from sediment using hydraulic jets, then transporting these nodules via pipe to a ship on the surface using hydraulic or mechanical lifting systems.

The quality of nodule resources and relative simplicity of their collection have made nodules the focus of most exploration efforts. Of the 31 exploration contracts in international waters granted by the ISA, 19 target nodules while only 7 and 5 target polymetallic sulfide deposits and ferromanganese crusts, respectively.[476] Moreover, technology for harvesting nodules is approaching commercial readiness, with several Western operators having successfully tested collector vehicle systems.[477] Chinese state-owned companies, benefiting from coordinated support from industry partners, academic researchers, and national policy, intend to test similar technologies in international waters in 2025.[478]

Given current momentum, it seems likely that seafloor nodule collection will proceed with or without the United States. How soon this new frontier of metal resources manifests depends on many factors: technological performance and costs, the speed at which regulatory frameworks develop, government policy and scientific support, and the scope of demand induced by commodity markets. In any event, the United States should take proactive steps to leverage deep sea nodule resources. A fresh, ambitious policy approach could establish U.S. leadership in a strategic emerging technology sector and strengthen the national critical minerals supply picture in a single stroke.

WHY DO SEAFLOOR NODULES MATTER TO THE UNITED STATES?

The Clarion-Clipperton Fracture Zone (CCZ) contains the largest polymetallic nodule field in the world, with 21 billion dry metric tons (23 million tons) of nodules. With a typical mass composition of 31 percent manganese, 1.4 percent nickel, 1.2 percent copper, and 0.2 percent cobalt, the CCZ may hold 6 billion metric tons (6.6 tons) of manganese, 270 million metric tons (300 tons) of nickel, 230 million metric tons (250 tons) of copper, and 50 million metric tons (55 tons) of cobalt.[479] The CCZ alone thus exceeds known land-based cobalt and manganese resources—not reserves—by a factor of two, while holding nickel equivalent to over half of terrestrial resources. This quantity of minerals can produce up to 6.9 billion electric vehicle (EV) battery packs, more than enough to support the global transition to EVs a few times over.

Seafloor nodules thus offer a tantalizing opportunity to help the United States diversify four critical minerals supply chains at once while supporting domestic EV manufacturing. Chinese industry players currently dominate the value chain for cobalt, nickel, copper, and manganese. Around 70 percent of all global cobalt and nickel processing capacity operates in China, as does 44 percent of global copper refining.[480] Consequently, Chinese firms perform over 75 percent of global manufacturing of nickel manganese cobalt (NMC) cathode materials for EV batteries.[481] This poses substantial challenges to U.S. automakers seeking to meet domestic-sourcing criteria for the critical minerals mined and processed for use in batteries.[482] Manganese nodules could overturn this calculus, playing to U.S. strengths in artificial intelligence (AI), robotics, and oceanographic science while significantly expanding the country's access to manganese, nickel, copper, and cobalt. Conversely, failure to capitalize on the seafloor nodule opportunity may funnel collected minerals to ascendant Chinese battery technology firms, further cementing their supply chain dominance.

In addition, recent economic assessments suggest nodule operations can produce minerals at attractively low prices relative to conventional mining.[483] Seafloor nodules could consequently support low-cost battery mineral feedstocks, increasing the economic competitiveness of downstream processing and battery manufacturing and helping reduce the costs of transitioning to clean energy and transportation technologies.

Some challenge whether United States should consider harvesting seabed minerals at all, citing environmental concerns.[484] Nodule collector vehicles will likely disturb the upper 5–15 centimeters (2–6 inches) of seafloor sediment and create a localized sediment cloud while introducing noise and light, disrupting organisms living on or around the nodules.[485] But considering the higher and harder-to-manage environmental impacts of conventional mining on land, policies focused on nodules actually offer the United States an opportunity to help reduce the long-term environmental impact of global mineral supply chains.

Seafloor nodule operations inherently allow for reduced environmental impacts, continuous improvement in impacts management, and strong enforcement of regulations. First, coproduction of four critical minerals from a single seafloor nodule area yields environmental benefits by obviating the need for multiple different mining projects on land. In contrast to land-based mining where operators must clear and excavate a sizeable area of the targeted landscape before extracting a single ton of ore, impacts from robotic collection are incremental, allowing for ongoing improvements that reduce the footprint of subsequent operations. The technology for nodule harvesting is still young, and successive generations of equipment may achieve better environmental protection—by selectively harvesting nodules with AI assistance, for instance, or by muffling underwater noise from the pipe system lifting nodules to the surface vessel.[486] Ship-based operations, which operate far from human settlements, also offer advantages for regulators seeking to monitor or audit nodule collection activities. Overall, a holistic comparison of seafloor nodules relative to land-based minerals arguably increases the value of nodule resources from a policy perspective.

THE CURRENT INTERNATIONAL STATE OF PLAY

To develop an effective U.S. strategy for polymetallic nodules, policymakers will need to consider the current state of play in international regulations and governance, seafloor exploration and pilot testing, and parallel policy efforts advanced by other nations.

For international waters beyond countries' sovereign EEZs, UNCLOS established the International Seabed Authority to oversee and regulate commercial activities on the seafloor. The United States is the sole major country not party to UNCLOS and only participates in the ISA as an observer, unable to sponsor applications for exploration or exploitation contracts. While the Deep Seabed Hard Mineral Resources Act of 1979 theoretically tasks the National Oceanic and Atmospheric Administration (NOAA) with issuing licenses for seabed minerals beyond the U.S. EEZ—with Lockheed Martin holding rights to two areas of the CCZ since 1984—the likelihood of intense diplomatic and political backlash strongly discourages both NOAA and private sector actors from unilaterally claiming seafloor minerals outside the UNCLOS framework.[487]

Since commencing operations in 1996, the ISA has seen seabed mineral collection evolve from a vague hypothetical into a technologically viable emerging industry. Although the authority has implemented a permitting framework for seabed mineral exploration, the original 2020 target date for finalization of a permitting system for mineral exploitation has come and gone, delaying completion of the ISA's Mining Code.[488] The 29th session of the ISA, in mid-2024, saw the first complete reading of the draft regulations on exploitation. The organization seeks to complete its regulatory framework by mid-2025, subsequently requiring approval by consensus from the rotating 36-member ISA Council and the 170 members of the full ISA Assembly.[489] As per the text of UNCLOS, "consensus" means "the absence of any formal objection," meaning any council or assembly member can stall codification of ISA regulations.[490] With a growing list of 32 member countries calling for a precautionary pause or moratorium on deep sea mining until the ISA formalizes robust regulations—or until researchers can study potential environmental impacts more extensively—this approval process faces a real risk of failing to reach consensus over the next few years.[491]

Still, U.S. policymakers would be remiss in deprioritizing seafloor minerals on account of the ISA's sluggishness. The authority only governs mineral collection on the international seafloor, leaving countries free to pursue activities within their own EEZ boundaries. While some countries have committed to applying the ISA's Mining Code to guide permitting of mineral exploitation activities in their own EEZs, this is voluntary and not universal practice. As such, commercial seabed mining in territorial waters may potentially outpace policy developments at the ISA.

ISA regulations are also steadily progressing as underwater technology improves and scientific knowledge accumulates. The newly elected secretary general of the ISA, Leticia Carvalho, has expressed determination to maintain the authority's neutrality as a decisionmaking forum, as well as delivering a consensus by 2025 on how member states will approach finalizing the Mining Code.[492] Such efforts will gradually facilitate consensus building at the ISA sooner or later. Given the current pace of resource mapping and collector-system testing, a finalized permitting framework will likely see nodule collection efforts move rapidly to full-scale production, spearheaded by both Chinese and Western companies.

Indeed, Chinese operators are moving ambitiously to develop international seabed resources, aiming to become first movers once regulations fall into place. Overlooked by most media reporting and unhindered by environmental activists, Chinese research institutions, state-owned enterprises, and private sector partners are investing significant efforts into resource surveys and technology development for the exploitation of the three main seafloor mineral resource types. This network of corporations and research centers can fully study, finance, build, and operate all components of a maritime minerals industry, including robots, specialized vessels, and onshore ore processing facilities. Many of these deep sea mapping, monitoring, and robotics platforms possess military dual-use value for submarine, anti-submarine, and reconnaissance operations and for targeting seafloor pipelines and cables.[493]

China's multi-sector deep sea mining industry efforts, which enjoy clear national policy backing, stand in stark contrast with the United States' nonexistent seafloor minerals strategy. The Chinese state-owned firms Beijing Pioneer and China Minmetals Corporation plan to conduct nodule collector field tests in 2025.[494] Beijing is also the largest financial contributor to the ISA and plays an active role in negotiations, vocally advocating for a firm pathway toward finalizing the ISA's regulations on exploitation.[495] Overall, China holds five ISA exploration contracts, the most of any single country: three for nodules, one for ferromanganese crust, and one for polymetallic sulfides.[496]

Of the remaining 26 ISA contracts, U.S. partners through NATO and the Minerals Security Partnership hold a total of 15 contracts, including 8 for nodules. However, only Belgium, India, Japan, and South Korea have shown clear interest in developing minerals on the international seafloor, while the remaining 4 contracts for nodules belong to U.S. allies—Germany, the United Kingdom, and France—that have expressed some opposition to near-term commercial-scale nodule collection.[497] Importantly, small island states hold an additional 5 nodule exploration contracts, 4 of which— Cook Islands, Kiribati, Nauru, and Tonga—partner with U.S.- and Canada-based companies.

Overall, current international nodule exploration efforts remain competitive with Chinese programs from a technological, business, and resource-quantity perspective but exhibit more vulnerabilities to activist opposition and shifting national policies. The selection of players that will emerge as early leaders in nodule collection and related deep sea technologies thus remains an open question—one the United States still has opportunities to shape.

RECOMMENDATIONS

In contrast to China's efforts—characterized by significant capital investment, active ISA participation, and the tight integration of government, academic, and industry work—the United States essentially does not possess a national policy strategy on seafloor minerals. Over the past three decades, the federal government has not moved meaningfully to update and modernize regulations governing offshore minerals and has undertaken only sporadic efforts to research seabed mining. Broader underinvestment in critical minerals supply chains, such as the current lack of any operating domestic nickel- or cobalt-refining facilities, also inhibits U.S. competitiveness.

One might conclude that Washington should seek to ratify UNCLOS to gain access to polymetallic nodules and other resources and secure stronger legal protections for the United States' extended continental shelf.[498] However, ISA membership will only allow U.S. access to nodules over the long term. The immediate benefit of ratifying UNCLOS would be geopolitical: The United States has endeavored to protect freedom of maritime navigation with neither the ability to hold leadership positions at UNCLOS institutions nor access to the UNCLOS dispute settlement framework, while also enduring accusations of hypocrisy for enforcing a treaty it is not party to.[499] At the ISA, it would be playing catch-up, with limited ability to help formulate initial ISA Mining Code rules, regulations, and procedures. Meanwhile, the United States would require significant lead time to directly sponsor new applications that secure contracts to explore and exploit international seabed resources.

While accession to UNCLOS would greatly strengthen U.S. maritime interests, political constraints pose obstacles not only to Senate ratification but also to seafloor mineral policy in general. Some conservative policymakers have consistently blocked UNCLOS ratification, arguing that participation would require the United States to relinquish sovereign powers to international institutions.[500] Meanwhile, the State of Hawaii recently joined California, Oregon, and Washington in enacting a moratorium on mineral exploitation in state coastal waters (which typically extend to 3 nautical miles offshore).[501] These moratoria pose minor implications for critical minerals, as the vast majority of hardrock mineral deposits are located further offshore. Yet these state policies highlight interest from other, liberal political coalitions in opposing ocean mineral exploration.

Such political tripwires pose difficulties for developing seafloor mineral policies that can secure bipartisan support. However, measures to establish downstream U.S. access to polymetallic nodules, coupled with support for offshore resource mapping, deep sea technology development, and onshore nodule processing, may be able to attract broader political buy-in.

In the near term, the United States should strive to position itself as a promising destination for processing nodules and incorporating refined minerals into the domestic battery supply chain. One of the advantages of nodule resources is that the United States could receive nodule materials without holding ISA contracts of its own, as long as domestic facilities can refine the metals competitively. Inaction will allow Chinese metallurgical industries to master nodule processing, gaining a monopoly even on nodules collected by non-Chinese operations.

However, the United States can pursue several actions to gain advantages in this sector.

- **Clarify existing policies to better support development of a domestic processing sector.** The domestic critical minerals component of the IRA's 30D EV tax credit provides a strong incentive for battery and EV manufacturers to source eligible metals mined or processed domestically or by free trade agreement (FTA) partners. Yet neither the IRA nor subsequent Department of Treasury guidance has addressed the eligibility of minerals sourced from the international seafloor.[502] Clarification that collected nodules brought to the United States and FTA partners as the first port of call would count toward the Treasury Department's Traced Qualifying Value Add Test for the 30D credit would incentivize bringing harvested nodules to the United States.

- **Fund research and pilot projects for nodule processing.** Mineral recovery from nodules will require different processing steps and workflows, opening a window for new market entry. Onshore processing of nodules will likely utilize many smelting and chemical-leach processes either in wide practice or with previous commercial-scale precedent.[503] But polymetallic nodules are not a standard feed material for metallurgical refineries and will require unique process flowsheets, of which industry actors are currently engaged in lab and pilot-scale testing.[504] Full process demonstration and product quality verification help operators prove their viability and move toward full-scale projects. As such, nodule-processing pilot and demonstration projects could benefit greatly from Department of Energy and Department of Defense awards through the IRA's 48C credit, the Defense Production Act Title III office, and the Loan Programs Office.[505]

But making the United States a destination for harvested nodules depends on the overall economic viability of domestic mineral-processing operations. As such, broader efforts to improve the competitiveness of domestic processing and refining—through affordable energy inputs, permitting reforms, and policy interventions to combat price volatility—will also help make the United States more attractive as a host country for downstream industry.

- **Incorporate polymetallic nodule collection into the United States' international critical minerals strategy.** The Department of State should proactively identify opportunities to expand the Minerals Security Partnership (MSP) to countries that ambitiously pursue nodule collection. Diplomatic engagement, technical assistance, and facilitated financing could strengthen early efforts by states such as the Cook Islands, Jamaica, or Nauru to develop nodule resources. Such partner countries could even consider sponsoring U.S. firms to explore or exploit nodules on the international seabed. Policymakers can similarly coordinate with existing MSP partners such as India, Japan, or South Korea that are advancing EEZ seafloor mineral exploration efforts. Robust diplomatic efforts can not only help diversify supply chains but also create new avenues for U.S. access to nodules.

- **Deepen scientific understanding of seafloor minerals to improve public opinion, study offshore mineral potential, and advance better environmental and economic outcomes.** Infrequent policy attention toward offshore hardrock minerals over the past few decades has left much of the extended continental shelf unexplored by modern oceanographic sensing techniques.[506] Surveying efforts can support and expand upon the efforts of the Bureau of Ocean Energy Management and U.S. Geological Survey to catalog a National Offshore Critical Mineral Inventory, potentially identifying promising new U.S. offshore deposits through mapping and characterization. Other research should seek to quantify industry-related questions such as life-cycle environmental impacts and economic feasibility. Many of these activities can inform longer-term modernization of federal regulations for leasing and permitting offshore minerals. Federal agencies can also help engage environmental activists' concerns through reports that review the scientific literature on potential impacts from seafloor nodule collection, track advances in collection system technologies, and highlight remaining questions and opportunities for impact mitigation.

Combined, these actions can position the United States far more competitively on polymetallic nodules, a substantial new critical minerals resource with the most promise for sustainable and economic development among known seabed mineral deposit types. Nodule collection can allow the United States to dramatically diversify four key metal supply chains for the clean energy transition while simultaneously reducing environmental impacts associated with these raw materials. Decisive policy efforts in this area can also strengthen nationally important industries such as robotics and deep sea capabilities, enhance U.S. overseas partnerships, and elevate U.S. leadership in strategic international forums. A new Congress and executive administration should not hesitate to integrate seafloor nodules into the nation's critical minerals strategy.

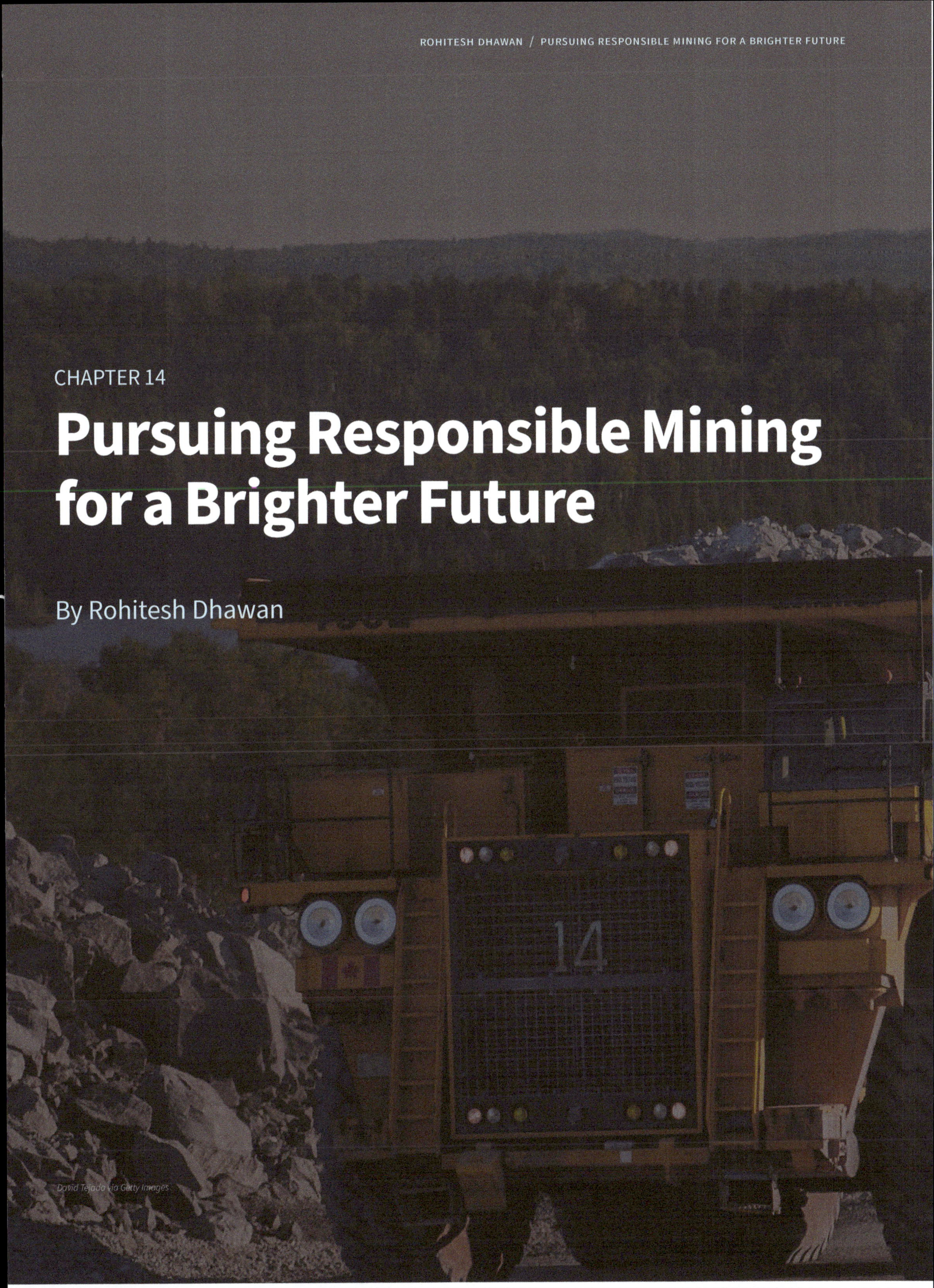

CHAPTER 14

Pursuing Responsible Mining for a Brighter Future

By Rohitesh Dhawan

David Tejada via Getty Images

Leading mining practices today can ensure that critical minerals are produced with minimal freshwater use, a small footprint on land, few if any local or global air pollutants, and a net-positive impact on plants and animals.

INTRODUCTION

Mining is an ancient industry, with evidence of copper piping used in Egypt from 2500 BC and other more basic uses of metals and minerals stretching back many centuries further. Yet modern and responsible mines bear little resemblance to the rudimentary and often damaging practices of bygone eras. Leading mining practices today can ensure that critical minerals are produced with minimal freshwater use, a small footprint on land, few if any local or global air pollutants, and a net-positive impact on plants and animals. In addition, such projects enjoy the support of local communities and Indigenous peoples; contribute to local and regional economic development through jobs, taxes, and infrastructure; and ensure the health and safety of workers and neighbors. Many such examples exist in the United States and around the world.

Yet for every example of responsible mining, there are numerous examples of irresponsible practices, past and present. Polluted lands and water, a depleted state of nature, and long-term health impacts on workers and communities are a reality for many mining communities today. At least 23 million people are estimated to live on floodplains contaminated by potentially harmful concentrations of toxic waste from metal mining.[507] Mining activities have also often been linked to human rights violations, conflict, and corruption.

Any industry—particularly one as diverse as mining, which has an estimated 25,000 companies operating in 140 countries globally—has a range of responsible and irresponsible actors.[508] The reputation of the overall industry is shaped disproportionately by the actions of the less responsible, setting in motion a vicious cycle that continues to create the space for irresponsible mining. A poor reputation for mining overall means less public and community support for new mines, which discourages responsible companies from seeking to operate, especially as developing a new mine can require billions of dollars before the first ore is extracted. Land is thus left open for illegal and irresponsible operators to take over, often undetected—as mining activities often take place in remote locations, governments may struggle to enforce local laws and regulations (where they exist), particularly in countries where the state has less capacity.

While it may be impossible to eliminate all irresponsible mining everywhere, it is possible to make responsible mining the norm in the United States and its allied countries. Doing so requires action on three fronts: regulatory measures, voluntary standards, and market mechanisms. Each is discussed below, and regulatory measures are covered in more detail in Chapter 10 by Morgan Bazilian and Gregory Wischer.

REGULATORY MEASURES

Mining is by nature a highly regulated industry in the United States and in most countries. Aside from mining-specific laws, a range of environmental, social, and economic legislation typically applies, such as the National Environment Policy Act and the Clean Water Act. Both are necessary and potentially useful, as they provide the requisite infrastructure through which to ensure responsible mining practices. Yet there are several challenges regarding the application of regulatory measures to mining. A new mine can require hundreds—and in some cases, thousands—of permitting obligations, take years of effort prior to approval, and put significant financial investment at risk. The substance of these requirements, however, are essential social and environmental safeguards that should be preserved.

There is industry and government consensus that there is excessive bureaucracy, uncertainty, and complexity in current legislation and its application in the United States. Globally, there is data for 127 mines that have the International Council on Mining and Metals (ICMM) in 2003.[516] Adopting these principles is a condition of membership for the organization's 24 mining and metals companies, which collectively make up approximately one-third of the global industry across 650 sites in over 50 countries.[517]

Other notable examples include the Towards Sustainable Mining program developed by the Mining Association of Canada, which has now been adopted in several other countries; the Copper Mark, which brings together the mining and business components across the value chain through a robust assurance system; and the Responsible Gold Mining Principles developed by the World Gold Council.[518] Each of these standards aims to promote continuous improvements in responsible mining practices for the facilities that voluntarily elect to adopt them. There are important differences across them, however, as they have varying levels of breadth (number of topics covered), depth (the specificity of performance requirements), governance (who decides what is or is not covered by the standard and how claims are assessed), and assurance systems (how third-party verifiers are required to substantiate claims made against the standard).

In addition to "upstream" or mining-specific standards, various "downstream" or product-specific standards have also been developed, including, among others, ResponsibleSteel and the Aluminium Stewardship Initiative.[519] These standards are focused on individual commodities and therefore seek to cover all stages of the value chain, from the initial extraction of the ore through the various processing stages to the development of the final product. They are best thought of as "product-back" standards that contrast with the upstream systems, which can be seen as "mining-forward."

The challenge today is that the mining companies that self-identify as being responsible often apply multiple voluntary standards, which adds significant costs and audit burdens in addition to causing confusion and complexity for stakeholders. On the other hand, large portions of the 25,000-strong overall mining industry adopt no voluntary standards at all, thus escaping the scrutiny and discipline that come with the disclosure requirements associated with these standards.

There are ways to maximize the effectiveness of voluntary standards as a tool to drive more responsible mining. Primary among these methods is support for the consolidation of existing standards. The current landscape of voluntary standards is crowded, confusing, and cumbersome for both the companies that apply them and the stakeholders who rely on them. ICMM, the Mining Association of Canada, the Copper Mark, and the World Gold Council are already in the process of consolidating their standards into one global system with independent, multistakeholder oversight.[520] Additional efforts should be supported by resisting any attempts to create further standards and incorporating—where relevant and appropriate—compliance with high-quality voluntary standards, as the European Union refers to in its Critical Raw Materials Act.[521] Conformity with high-quality voluntary standards could also be made a condition of public procurement to help scale up adoption.

Voluntary and regulatory mechanisms can be complementary tools to drive the widespread adoption of responsible mining practices. However, there is a third category of actions that is necessary to enable the maximum uptake of such practices: market mechanisms.

MARKET MECHANISMS

While metals and minerals are bought and sold in a variety of ways, including direct sales from producers to customers (who may still transform or on-sell the product), commodity exchanges such as the Chicago Mercantile Exchange or London Metal Exchange (LME), and illegally or "off-the-books" trading, particularly in the case of precious metals such as gold. By and large, individual tons of a particular commodity of a specified grade are indistinguishable from each other and have the same essential chemical and physical properties, irrespective of whether they were produced using responsible or irresponsible methods.

This fungibility of metals gives rise to a major challenge in making responsible mining the norm. Put bluntly, as long as there are both a clean-but-expensive way of mining and a cheap-but-dirty way, many will choose the latter since the customer cannot usually tell the difference based purely on the attributes of the product. This is different from, for example, farm produce, for come online since 2002 and the discovery-to-production process has taken an average of 15.7 years.[509] The United States is significantly behind the global average. A recent study by S&P Global on behalf of the U.S. National Mining Association showed that the country has the second-longest mine development times in the world, at almost 29 years on average from first discovery to first production—second only to Zambia, at 34 years.[510] The rate of slow mine development is problematic. The expected demand for copper alone is likely to result in the need for between 35 and 195 large new copper mines over the next 30 years.[511]

How, then, can the permitting process be streamlined while maintaining strong social and environmental safeguards? There are three important methods the U.S. government can consider, drawing from the experience of other jurisdictions.

1. **Implement a "one-stop-shop" mechanism.** While the form of this mechanism will no doubt vary from country to country, it would not only cut through the complex systems of regulations but, crucially, could also ensure authorities strike the right balance between fast-tracking applications and implementing social and environmental safeguards. For instance, the EU Critical Raw Materials Act (CRMA) allows for "strategic projects" (defined as those that extract identified strategic raw materials) to benefit from a streamlined and predictable process to help project developers navigate the national permitting systems.[512] This approach is designed to expedite permits through a single authority and administrative system, while at the same time ensuring that environmental review, public participation, and sustainability remain embedded in the approval process. This is difficult—if not impossible—to achieve without a single point of contact. Other countries that already have or are considering similar single-point systems include Chile, Oman, and Saudi Arabia.

2. **Set some "non-negotiable" elements while allowing flexibility in others, ensuring that expedited permitting can go hand in hand with high standards.** For instance, the CRMA's non-negotiable requirements include compliance with the conditions set in the directives on habitats (92/43/EEC), water (2000/60/EC), and birds (2009/147/EC), as well as in the European Union's recent Nature Restoration Law.[513] If it meets these criteria, a strategic project can benefit from the expedited permitting process even if it has residual negative impacts in other areas. Another way the CRMA strikes this balance is to exclude the time the project developer takes to complete the Environmental Impact Assessment (EIA) from the rapid-approval time window. This exclusion is because the EIA is the responsibility of the project developer, and it ensures that the EIA process is not unduly rushed and that steps are not skipped. With this protection in place, the act further streamlines the process by requiring that a decision on "screening" under the EIA directive (2011/92/EU) should be made within 30 days, rather than a previous timeframe of 90 days, and that the maximum time for public participation in the EIA process should not exceed 85 days.[514]

3. **Coordinate, cross-link, or "nest" requirements of different acts within one another.** For example, in Western Australia, the Mining Act and the Environmental Protection Act are coordinated. In addition, the country's Department of Mines and Petroleum has a memorandum of understanding with its Environmental Protection Authority that delegates certain tasks to the former. Various conditions—in accordance with various laws—are appended to the licenses obtained under the Mining Act. Furthermore, the Mining Act links the granting of mining rights to provisions on how the environment should be handled. This effectively consolidates all environmental provisions within the legislation, in turn ensuring a streamlined and administratively efficient process while minimizing gaps and overlaps that could lead to poor environmental or social outcomes.[515]

But regulatory reform is only one way to encourage responsible mining. An often more powerful lever is the effect of voluntary standards.

VOLUNTARY STANDARDS

Over the past two decades, numerous voluntary standards for responsible mining have been developed, mostly initiated by the mining industry itself. One of the earliest examples is the Mining Principles put forth by which irresponsible practices such as excessive use of fertilizer could be ascertained through visual or chemical testing.

This perverse incentive for irresponsible practices is made worse by the lack of a well-developed mechanism for pricing metals differently based on their provenance or quality of production. This is most acute in the case of commodity trading platforms such as the LME, which essentially uses a single reference price for a commodity. The LME has responsible sourcing requirements traders must satisfy before being able to do business on the platform, theoretically setting a "floor" of responsible mining practices.

However, beyond this, there are few if any mechanisms currently available on exchanges such as the LME to allow commodities to be priced higher if they have been produced more responsibly or, conversely, priced lower if they lack certain assurances on their provenance or performance on sustainability criteria.[522] This has come into sharp focus in the case of nickel, where the divergence between responsible and irresponsible practices can be quite large. In particular, the greenhouse gas emissions associated with production can vary by orders of magnitude even within the same class of nickel, which is based on method of production. Despite a growing call, particularly from Western producers, for the LME to introduce differential pricing for "green nickel," the exchange has not yet moved to create such tiered pricing, citing a lack of liquidity and critical mass and highlighting the risk of the market not working efficiently if contracts were to be subdivided further.[523]

Part of the challenge lies in defining "green." There is as much debate about what metrics should be considered (should it only be greenhouse gas emissions and other environmental factors, or should it also include social performance?) as what data would be used and how it would be collected and assured (would site-level Scope 1 and 2 emissions suffice, or would it need to include upstream and downstream Scope 3 emissions, too?).[524] The lack of clarity around these issues has curtailed the development of tiered pricing mechanisms.

As a result, operators that do not appear to follow any voluntary codes have flooded the market with metal produced amid credible allegations of environmental destruction. This has caused the shuttering of large nickel companies that do follow voluntary responsible mining practices, such as the Nickel West operations and the West Musgrave project in Australia.[525] This trend can be expected to continue unless there are efforts made to reward those following responsible practices through some form of price mechanism.

These risks to market efficiency are real and need careful work and attention—especially given the even bigger challenge posed by not having clear market mechanisms to incentivize responsible production. The solution may not be merely to institute a "green premium," but could include provisions such as tiered pricing in public procurement or other requirements for market access. However, these proposals all depend on having clear and commonly accepted definitions of "green" and "responsible" production, hence the need for consolidation and harmonization of voluntary standards, as previously mentioned, and clarity on how these fit with regulatory requirements.

RECOMMENDATIONS

Based on the analysis of regulatory, voluntary, and market mechanisms to encourage responsible practices, the U.S. government should:

1. **Create a national panel on mining.** This idea emanates from the Towards Sustainable Mining (TSM) voluntary scheme developed by the Mining Association of Canada. The TSM Community of Interest Panel is an independent group of stakeholders, communities, and nongovernmental organizations that oversees the implementation of the standards program. Consisting of 12 to 15 members, the panel is designed to provide guidance to auditors and implementers on the issues of particular importance and resonance in the local context.[526]

   The United States could benefit from importing and adapting this panel structure, even if it currently does not and may never adopt the TSM program. Such panels, which include cross-stakeholder dialogue leading to improved understanding and trust building, are useful and important in their own right, even if not anchored to a voluntary scheme. A U.S. version could help shape domestic regulation, inform policy discussions, and resolve conflicts. And it may not need to be a single body but could instead involve multiple panels for different regions. Engaging the National Mining Association to explore the feasibility of such a mechanism would be an appropriate early step.

2. **Incorporate voluntary responsible mining standards into trade agreements.** Over the past 10–20 years, governments in both producing and consuming countries have increasingly incorporated voluntary sustainability standards into trade policy instruments—particularly free trade agreements (FTAs)—to address environmental, social, and economic concerns regarding commodity production and trade. Before 2018, there were only 17 FTAs worldwide referring to Voluntary Sustainability Standards (VSS) across different sectors. However, almost half of all new agreements over the following five years referred to them in some way.[527]

   There is no notable example of voluntary responsible mining standards being included in FTAs, and research on commodity-specific standards being included in such instruments is limited and available only for certain agricultural products.[528] Yet including such standards in trade agreements would be uniquely suited for the United States, given its importance in the global trade ecosystem and its ability to use trade to influence practices both domestically and abroad.

   There are different ways in which voluntary mining standards could be incorporated into trade agreements. These can range from loose to broad measures, including exchanging information on the provenance and socio-environmental performance of commodity production and promoting cooperation in areas such as labeling, to more specific aspects, such as giving preferential treatment to products that are certified under particular schemes. The most notable example of this from another sector is the European Free Trade Association (EFTA)-Indonesia Comprehensive Economic Partnership Agreement, whereby palm oil imported into Switzerland from Indonesia benefits from preferential tariff treatment if producers comply with one of three specified voluntary schemes.[529]

   It would not necessarily be advisable to include specific criteria in the case of mining and metals given the rapidly changing landscape of responsible mining standards, but such incentives could be considered by indicating the attributes schemes need to have rather than naming particular ones.

3. **Support the development of "green premium" instruments.** Venues where commodities are traded, such as the Chicago Mercantile Exchange, face legitimate barriers to developing instruments that would offer suppliers of responsibly produced metals a "green premium." At the same time, the industry's ability to voluntarily develop such mechanisms is limited by antitrust laws and regulations, which prohibit coordination action by producers who might collude to manipulate or otherwise alter prices. This gives the U.S. government the opportunity to explore various mechanisms for supporting the development of green premiums.

   First, U.S. public entities can participate in buyers' clubs that favor and reward responsible producers of mineral-based products. A notable example is the Sustainable Steel Buyers Platform, which aims to pool demand for responsible steel from different buyers, thus giving impetus and scale to the producers of these commodities.[530] Having a deliberate focus to participate in such initiatives can ensure that through their sizeable purchases of minerals, U.S. public entities help grow the market for responsibly produced products.

   Second, the United States can facilitate or lead the development of a multilateral pilot initiative among leading Western commodity exchanges to assess the feasibility of pooling trading activities in a way that provides large enough markets for contracts to be priced based on sustainability criteria. The Chicago Mercantile Exchange and the London Metal Exchange would be the most obvious participants in such a pilot. This could be funded by the U.S. government, thereby overcoming one of the main barriers holding back the development of such instruments: the lack of liquidity in any individual exchange.

CONCLUSION

For a segment of society that believes any form of extraction from the earth is by definition irresponsible, the term "responsible mining" seems an oxymoron. Yet mineral production can incorporate a wide range of responsible practices. Given the essential role of metals and minerals in modern society and the reality of the need to extract significantly greater quantities in the future, there is perhaps no more urgent question than how to make this production as responsible as possible.

The United States has an essential role to play in this regard, both domestically and abroad. The three most salient actions the U.S. government can take include establishing a national panel on responsible mining, incorporating voluntary standards into trade frameworks, and supporting the development of green-premium instruments.

CHAPTER 15

A Comprehensive U.S. Critical Minerals Plan

By Frank Fannon

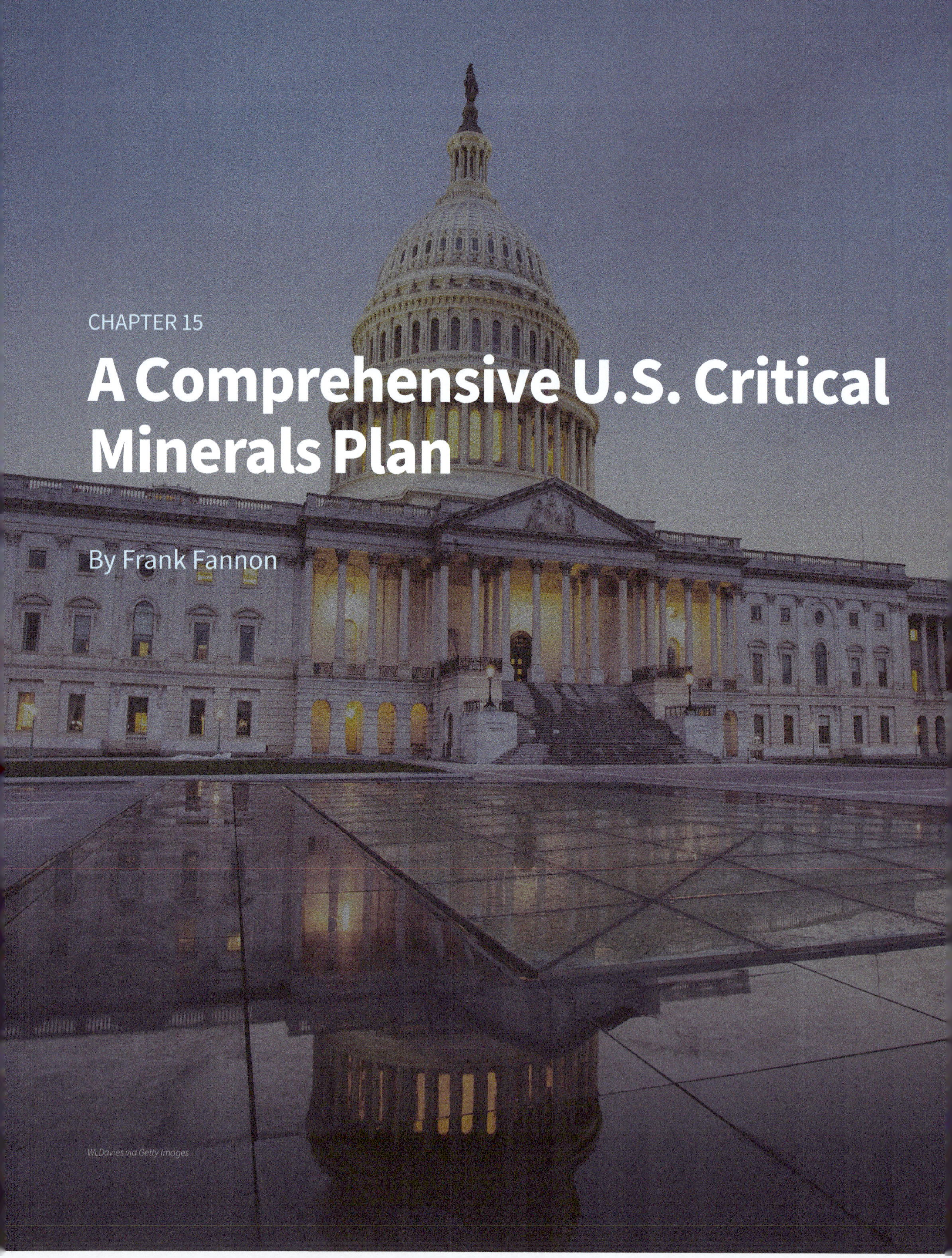

WLDavies via Getty Images

Washington cannot assume that traditional alliances or free trade agreement status indicate alignment with U.S. security interests.

The United States' dependence on foreign rivals, especially the People's Republic of China (PRC), for critical and strategic minerals presents a material vulnerability to its industrial, energy, and defense sectors. This vulnerability quietly developed over decades, only coming to public attention in 2010 after Beijing banned the export of rare earth elements (REEs) to Japan.[531] Although Washington has elevated critical minerals as a top security issue, the United States remains dependent on an increasingly adversarial China. This chapter explores the importance and urgency of the United States developing resilient and secure critical minerals supply chains and recommends a comprehensive strategy to do so.

LESSONS (NOT) LEARNED

The United States' dependency was decades in the making. The United States and Europe were both happy to offshore low-margin and oftentimes heavy and polluting industries to other nations and import the refined goods on a just-in-time basis. This system worked well for years. However, the Chinese Communist Party (CCP) has forced Washington to confront its out-of-sight, out-of-mind reliance on China multiple times across the past three presidential administrations. Presidents Barack Obama, Donald Trump, and Joe Biden each responded to China's provocative actions differently—but incompletely, given its continued dominance.

The Obama Years

President Obama assumed office as China dramatically increased its military spending, replaced Japan as the United States' largest foreign creditor, and in 2010 became the world's second-largest economy.[532] During his term, Beijing publicly launched the Belt and Road Initiative, a strategy of coercive investments designed in part to lock up natural resources for China.

In 2010, Japan detained a Chinese trawler captain after repeated instances of illegal fishing and ramming Japanese coast guard vessels.[533] In response, China temporarily banned the export of REEs to Japan and implied that it would impose new quotas. These actions skyrocketed REE prices by more than 400 percent, drawing condemnation from Washington and Brussels.[534] In response, the Obama administration (joined by Japan and the European Union) initiated a World Trade Organization (WTO) case against China in 2012, stating:

> Now, if China would simply let the market work on its own, we'd have no objections. But their policies currently are preventing that from happening. And they go against the very rules that China agreed to follow. Being able to manufacture advanced batteries and hybrid cars in America is too important for us to stand by and do nothing. We've got to take control of our energy future, and we can't let that energy industry take root in some other country because they were allowed to break the rules.[535]

Senator Chuck Schumer (D-NY) criticized the WTO case. He said, "There are faster ways to assert leverage on China than relying on the WTO, which could take years to resolve the case."[536] He instead called to restrict Chinese mining in the United States and limit World Bank funding of PRC mining projects.

After two years of deliberations, the WTO finally concluded in 2014 that China violated trade rules. The U.S. trade representative Michael Froman stated, "By upholding rules on fair access to raw materials, this decision is a win not only for the United States, but also for every nation that respects the principles of openness and fairness. Those principles are the pillars of the rules-based global trading system, and we must protect them vigilantly."[537]

Rather than restrict exports, China flooded the market with supplies, sending prices crashing. In less than a year after the United States won in the WTO, its only REE mine filed for bankruptcy protection.[538]

The two terms of the Obama administration witnessed a rising and more brazen and provocative China. In response, Washington looked to well-intended, albeit conventional, remedies such as diplomatic convenings and Brenton Woods–era institutions to help settle disputes. However, the United States was unable to limit China's rising dominance of the critical mineral sector.

This experience showed that conventional mechanisms are only effective when countries agree to the conventions themselves. Since China rejects them, the United States needs to consider an alternative approach to the traditional rules-based order.

The Trump Years

By contrast, the Trump administration viewed critical minerals as a proxy for U.S. economic and national security. In December 2017, Trump issued an executive order directing the Department of the Interior to develop a critical minerals list.[539] The resulting May 2018 report identified 35 minerals considered critical to the economic and national security interests of the United States, which informed the interagency's focus areas.[540] The report helped increase awareness of the PRC's control of critical mineral supply chains, beginning in emerging markets targeted by the Belt and Road Initiative.

The Department of State created new bilateral and multilateral initiatives, such as the Energy Resources Governance Initiative (ERGI), to lay the foundation for alternative investment channels to the PRC. For example, when the PRC sought to make major investments to secure critical minerals in Greenland, ERGI enabled the Bureau of Energy Resources to pursue a series of diplomatic engagements that successfully culminated in memorandums of understanding to support Greenland's geologic endowment and preference U.S. and allied investors.[541]

Overall, ERGI sought to elevate transparency, support mineral-producing countries, and eventually leverage the newly established U.S. International Development Finance Corporation (DFC) to provide seed capital to de-risk projects and attract otherwise reputation-sensitive investors.[542] However, the administration was unable to implement its permitting reform ideas, and although historic, the DFC made only one critical minerals investment.[543] Moreover, as the Trump administration utilized existing tools such as tariffs and diplomatic initiatives, it faced domestic permitting obstacles and lacked adequate financing tools to support many U.S.-based and U.S.-backed mining projects.

Even with new initiatives and approaches, the U.S. government must do more to support domestic and allied investment to develop a secure critical minerals supply chain. This may require establishing new government tools while rethinking and optimizing existing ones to fit the mission.

The Biden Years

President Biden sought to return the United States to a more conventional diplomatic position. He rescinded certain Trump-era energy sanctions (such as on the Nord Stream 2 pipeline), called for a review of China-directed tariffs, and immediately directed the administration to reenter the Paris climate accord. In further contrast to the Trump administration's focus on national security and defense, the Biden administration prioritized the climate crisis.[544]

Understanding that meeting its ambitious climate change and clean energy goals would require an exponential increase in critical minerals, President Biden issued Executive Order 14017 in February 2021, which mandated comprehensive reviews of supply chains across the U.S. government.[545] The Department of the Interior issued a new critical minerals list, adding 15 new minerals, bringing the total to 50.[546] Meanwhile, the Department of Energy (DOE) released its own new critical materials list for energy.[547] The DOE material list rightly included copper as a "critical" mineral, but the USGS list excluded it, even though it is vital for every part of the modern U.S. economy—including infrastructure, clean energy technologies, electronics, and automotives—and the International Energy Agency has forecasted a copper shortage.[548]

The passage of the Bipartisan Infrastructure Bill and the Inflation Reduction Act (IRA) established new federal programs, incentives, and, importantly, billions of dollars in funding to support clean energy supply chains.[549] The Biden administration channeled much of those direct grants and loans into domestic minerals processing and clean-tech manufacturing facilities, which should improve domestic capacities in both. The IRA has been so effective in attracting clean-tech investment that it has alarmed European partners, who sought to force "concessions" from President Biden to allow EU companies to benefit from certain IRA subsidies.[550]

The Department of State's Minerals Security Partnership seeks to accelerate the development of a clean energy supply chain by convening governments and industry.[551] The department also leads the Partnership for Global Infrastructure and Investment, which aims to promote mining-related investment, such as the Lobito Corridor project.[552] In addition, the DFC increased its investment in Techmet, a technology metals company, from $25 million to $105 million but has not diversified equity-level investments in any other mining investors or operating companies.[553]

The Biden administration's prioritization of climate action above other issues has contributed to pragmatic but conflicting messages. The IRA prohibits U.S. taxpayer funds from going to a "foreign entity of concern," which covers firms controlled by China, Russia, North Korea, and Iran. However, China is by far the largest and least expensive critical minerals producer and clean-tech manufacturer in the world. To disqualify Chinese content from receiving taxpayer subsidies, as per the law, would increase prices—making electric vehicles unattractive to many American buyers. Therefore, the Department of the Treasury amended its rules in December 2023 to allow up to 25 percent of otherwise disqualified Chinese content to receive U.S. taxpayer subsidies under the IRA.[554]

The Biden administration's actions appear to have had an impact. Beijing went back to the 2010 playbook, announcing curbs on the export of gallium, germanium, graphite, antimony, and REE technology.[555] China furthermore flooded the market with cobalt, crashing prices and putting the United States' only cobalt development project into care and maintenance.[556]

Based on these experiences across administrations, developing a responsible and secure clean energy supply chain will require two things to be effective. First, the government needs to have absolute clarity of mission in recognizing the challenge the United States is trying to overcome. Second, there must be accountability to guard against mission creep and navigate the complexity and equities across the government. This will entail marshaling the United States' limited resources around the mission to optimize impact.

A U.S. CRITICAL MINERALS STRATEGY

China remains the world's dominant producer, processor, and buyer of critical minerals. The CCP continues to push domestic policies that artificially stimulate demand for its strategic sectors, mobilize state financing to influence market dynamics, and shirk environmental and human rights protections to produce commodities at the lowest cost. These are long-standing tactics.

The United States is still behind, but over the past three presidential administrations, it has learned a great deal, gained political support, and increased its tools and capabilities to develop a meaningful and comprehensive response. However, an effective strategy will require a blending of the Obama, Trump, and Biden administrations' approaches and will test domestic and international relations.

Mission Clarity

The United States should be clear about its objective. National security and climate change are both important and interrelated, but the U.S. government ultimately needs to prioritize one over the other. It is hard to imagine how the world can address climate change by increasing reliance on China, the world's super polluter, to produce inputs for clean energy technologies.[557] The United States can and should develop a responsible and secure critical minerals supply chain necessary for economic growth, defense, and clean energy. However, realizing such a goal while reducing reliance on CCP-backed industry will be more expensive.

The United States has repeatedly tried to strike a middle road by partnering with China on climate change but holding firm on core principles such as human rights and environmental standards in critical minerals supply chains. The CCP has rejected such attempts. Rather the PRC rejects the primacy of addressing climate change and instead views the matter as a core point of contention within the U.S.-China bilateral relationship.[558]

Although disappointing, the CCP's approach is rational from its perspective. The party-state is motivated first to advance its interests and second to increase its leverage or control over the United States and the rest of the world. By dominating the critical minerals supply chain, China forces the United States to increase its dependence on its adversary—and therefore forces Washington to question its security positioning.

The U.S. government should be clear in its mission to develop secure and responsible critical minerals supply chains. Successive administrations have repeatedly recognized that China dominates the production and refining of critical minerals—and thus also the defense industry and clean tech. To contest this threat, the White House should explicitly articulate its intention to develop secure supply chains and phase down reliance on China—not to provoke hostile powers but to galvanize interested stakeholders. Clear and unequivocal goal setting will signal to partner governments, resource-rich countries, and investors that the United States is on the field.

Accountability and Coordination

The U.S. government should have a single point of accountability to oversee and coordinate the administration's multiple lines of effort.

The IRA and Bipartisan Infrastructure Law provided historic levels of federal funding and new programs to develop clean energy and critical minerals supply chains. This explosion in funding coincided with the dramatic expansion of agencies working on critical minerals. Under the Biden administration, when counting, 15 federal agencies claim a meaningful role in U.S. critical minerals policy. While the increased interest is a positive development, the lack of clear oversight and management of such a complex set of issues can lead to inefficiencies or agencies working at cross-purposes, which may frustrate the mission.

To remedy this, the U.S. government should appoint or designate a special presidential coordinator at a minimum of an ambassadorial level to manage the critical minerals portfolio. This will be a difficult but necessary role: Although every federal agency or cabinet member may have an interest, some interests are more consequential to achieving the mission than others.

Furthermore, this special coordinator should align U.S. policies to address the country's current pacing challenge. As mentioned, the IRA has accelerated domestic clean energy-related manufacturing. However, the United States has not taken sufficient action to increase supplies of the critical mineral inputs needed to feed these new gigafactories and industrial facilities. It may only take 5 years to build a plant but some 15 years to turn a discovered resource into a producing mine.

Update Finance Tools

The U.S. government should update and integrate its mission into international finance tools. The country has just two such financing entities: the DFC and the Export-Import Bank (EXIM). Both should have a clear critical minerals mandate and be empowered to act upon it, as well as the flexibility and resources to respond to the challenges of today.

The DFC was designed to advance U.S. foreign policy, which is why the secretary of state serves as the chairman of the agency's board. Yet, as the name suggests, the DFC must also consider a "development" impact. However, the statute does not provide a framework for weighing or prioritizing these factors. The U.S. government should be clear about its goals and financing, particularly as mining is such a long-term endeavor.

The DFC's equity and debt tools are intended to catalyze private sector investments into key industries in emerging market countries. However, the White House's Office of Management and Budget, like the DFC itself, chooses to treat equity investments as if they were grants, which for accounting purposes are treated as a loss. Furthermore, when the equity investment realizes its returns, those funds are returned to the Department of the Treasury, not the DFC.[559] This accounting treatment significantly limits the agency's ability to make the requisite investments.

This scoring problem is a historical practice, not a statutory requirement. The White House could remedy the situation by issuing new scoring criteria but appears unwilling to take on that political fight without an express congressional mandate to do so. As such, Congress should provide that directive and make the United States' primary international finance tool appropriate for the realities of the market and geopolitical statecraft.

The DFC's investments should be both strategic and commercial. With those goals in mind, and to improve political support for an expanded remit, the DFC should prioritize investments in domestic companies. Currently, there is no preference to support U.S. companies with U.S. taxpayer dollars over foreign parties.

The EXIM Bank is the United States' export credit agency (ECA). The 90-year-old institution must compete against the 115 foreign ECAs around the world, especially from the PRC. In its 2019 reauthorization, Congress recognized the threat and strength of China's investments and directed EXIM to establish the China Transformational Exports Program (CTEP).[560] Through CTEP, EXIM gained greater flexibility to lend to projects focused on 10 strategic industries, including critical minerals. Congress should continue to build upon CTEP and further lower the domestic content requirements that constrain EXIM's lending authorities. The bank should also have the clear ability to provide debt financing at the company rather than project level. By providing company-level lines of credit, EXIM can empower U.S. companies to take advantage of strategic projects in real time.

The United States has two important international finance tools. The DFC and EXIM must be rightsized for the challenges of today.

Permitting Reform, at Long Last

The U.S. government has long been talking about, but doing little to improve, its permitting process. The federal permitting process has grown into a complex and uncertain process regardless of project type, whether related to a natural gas pipeline or solar power installation.[561]

Bipartisan members of Congress have advocated for permitting reform but have made little substantive progress over the years.[562] The exhaustive federal permitting process is a main obstacle to meeting the IRA's clean energy goals.[563] The IRA contains billions of dollars to develop clean energy networks, which will require the construction of electric transmission lines and improvements to the grid. According to Representative Scott Peters (D-CA), "The problem is that the average line is taking 10 years to build, but seven years of that is process."[564] The timeline is even worse for mining. According to an analysis by S&P Global, it takes an average of 29 years to turn a discovered resource into a mine in the United States, the second-longest mine development time after Zambia.[565]

The realization that today's exhaustive permitting process is undermining clean energy goals has helped to expand the parties calling for reform. In July 2024, Senators John Barrasso (R-WY) and Joe Manchin (I-WV) introduced the Energy Permitting Reform Act, which aims to start addressing some of these challenges. The bill proposes to improve certainty in decisionmaking by requiring a final agency decision within 150 days, reducing administrative steps, and providing clarity over the controversial Rosemont decision.[566] The bill avoids some of the more controversial proposals, according to some industry advocates, such as tightening standing requirements to legally challenge projects or proposals to increase community engagement.

Although the bill has secured strong bipartisan support, environmental opposition groups have rejected the legislation, arguing that only clean energy, not oil, gas, and mining, should share in the benefits of permitting reform.[567] The bill—which goes too far for some but not far enough for others—represents an incomplete but positive and needed step forward.

Permitting reform often includes difficult and long-standing issues, particularly concerning the history of mining in the western United States. Yet the federal government's failure to address permitting—in the meantime allowing the purchase of minerals known to be produced in a manner inconsistent with environmental protections, respect for human rights, or inclusion of local communities—is patently wrong. The United States should address this challenge head-on, especially if mining operations are to scale up to meet current and future clean energy targets.

Sticks

The IRA provided billions of U.S. taxpayer-funded dollars as "carrots" to incentivize investment in clean energy technologies. Many of these carrots take the form of tax credits that seek to reduce costs for consumers. While subsidies are tried-and-true measures that can affect consumer behavior, such carrots alone are insufficient to remedy China's critical minerals dominance.

According to the Federal Bureau of Investigation, "The Chinese government is seeking to become the world's greatest superpower through predatory lending and business practices, systematic theft of intellectual property, and brazen cyber intrusions."[568] Specifically, the Biden administration's supply chain report found that overreliance on China for critical minerals and materials posed national and economic security threats.[569]

It is hard to imagine a situation where China would allow the United States to out-subsidize and erode its dominant market share. After all, the CCP is the world's leader in economic statecraft, blending coercive domestic consumption, state investment in strategic industries, provincial support of local champions, and a willful blindness toward best-practice standards. And mining has been the bedrock of Chinese domestic industrial strategy and foreign policy for decades.

In response, the United States should take a more active and definitive role in countering the CCP's market-manipulating activities. Some businesses may argue that the U.S. government should go to great lengths to deal with the Chinese state and Chinese private sector separately. However, this is a fool's errand, as there is little distinction between the two. According to Stanford's Center on China's Economy and Institutions, a large share of China's economy operates in a gray zone of mixed or blended ownership: "The number of private owners with direct equity ties with the state almost tripled between 2000 and 2019, and those with indirect equity ties rose 50-fold. The analysis suggests that equity ties to the state may have aided, not constrained, the growth of China's private sector."[570]

The United States should take a much more realistic approach to address the threat it faces. To start, the United States should prohibit the use of taxpayer funds to subsidize Chinese technology or critical mineral interests. Furthermore, the U.S. government should require any company receiving taxpayer funds to certify that any imported or incorporated Chinese content or technology meets reporting standards.

The United States should also consider critical minerals and clean energy supply chains in light of today's new era of economic realpolitik. Washington cannot assume that traditional alliances or free trade agreement status indicate alignment with U.S. security interests. For example, European Commission president Ursula von der Leyen would like the European Union to qualify for IRA subsidies even though several European EV factories are owned by Chinese companies.[571] And in November 2022, German chancellor Olaf Scholz, together with the heads of Volkswagen and other companies, met with Chinese leader Xi Jinping in Beijing to boost business ties.[572] Such moves to increase dependence on a strategic threat weaken free nations' shared security.

Even as it works to strengthen traditional alliances, the United States should take a more pragmatic approach. U.S. and European officials have discussed creating a critical minerals buyers' club, but to be credible, club membership should be dependent on more rigorous criteria than just geography.[573]

CONCLUSION

In 2010, China banned the export of REEs to Japan. In so doing, the CCP fired the first, transformative salvo in an ongoing fight to leverage its critical minerals dominance to coerce, intimidate, and extort. The United States has since learned some valuable lessons. Successive administrations have tried multiple, albeit incremental, remedies aimed at encouraging China to behave responsibly and incentivizing U.S. and allied companies to reorient their consumption.

Although meaningful, these incremental tactics have not altered the CCP's strategy, and the United States has failed to develop secure supply chains. Building on the experience of the past three administrations, the United States should follow the above guidance to achieve its objectives.

U.S. leaders should recognize that, given the scale of the challenge, the federal government has a meaningful role to play. Yet, the country's comparative advantage lies instead in its dynamic and world-leading private sector. To that end, U.S. diplomacy and financial tools should be rightsized to achieve the mission. This rightsizing must also apply to domestic policy. Leaders must finally take on long-standing special interests to advance meaningful permitting policy reform.

The United States has been forced to engage in a new era of economic realpolitik. This awareness requires the United States to reconsider traditional alliances and partner relationships at a company or project level. Chinese companies have expanded and, in certain instances, entrenched themselves within traditional allies' commercial interests. As such, the U.S. government needs to guard against unintentionally supporting adversarial interests.

In his famous speech launching the Space Race, President John F. Kennedy asked Americans "to accept a firm commitment to a new course of action, a course which will last for many years and carry very heavy costs."[574] Kennedy's Apollo program was transformative for the United States' leadership in the world and led to innumerable technological innovations.

Transforming the United States' economic engine presents a challenge orders of magnitude greater than putting a man on the Moon. But while building secure clean energy and critical mineral supply chains will be neither easy nor inexpensive, it is increasingly vital.

CHAPTER 16

Conclusion

By Gracelin Baskaran and Duncan Wood

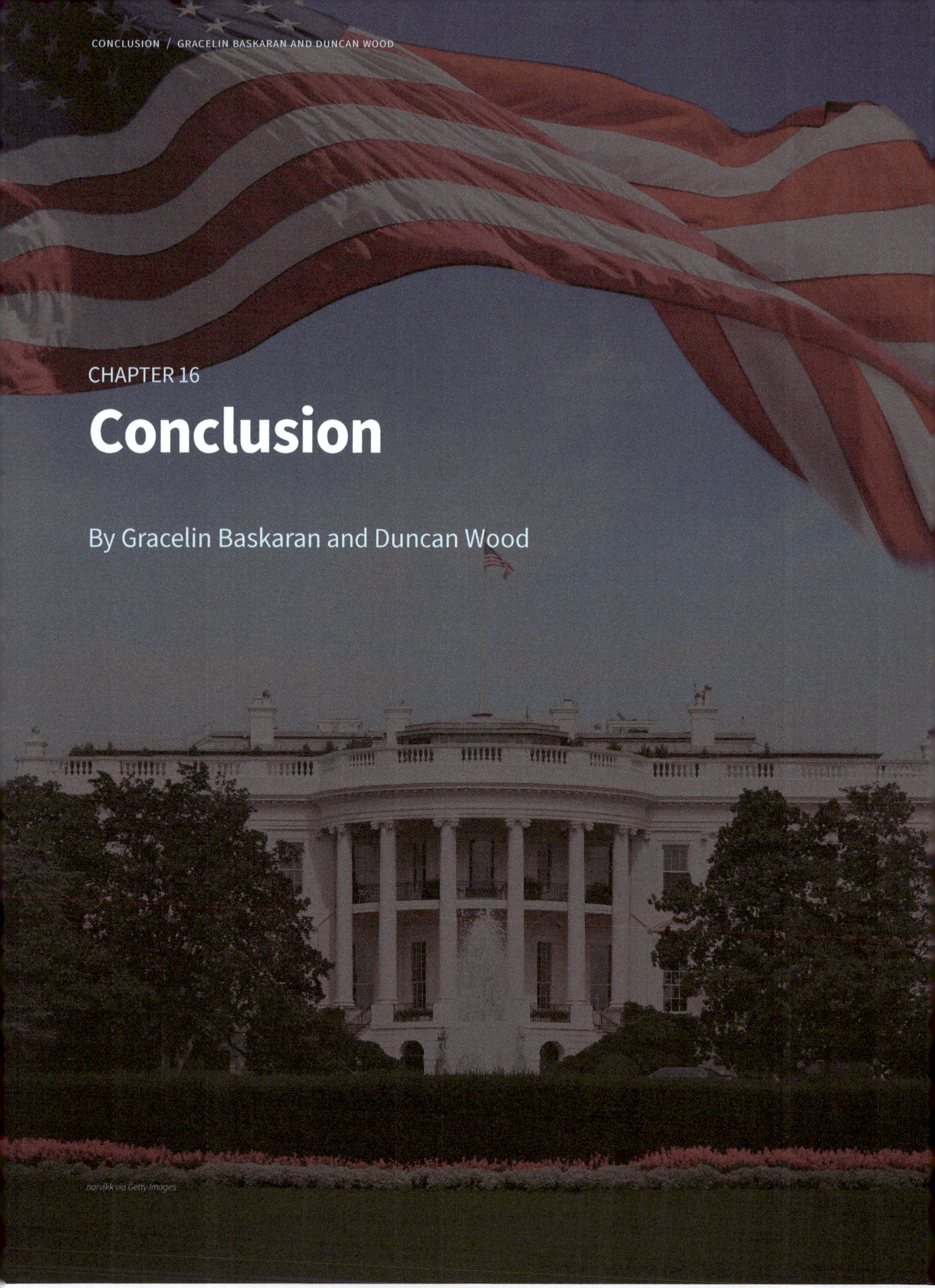

narvikk via Getty Images

This book has highlighted the cross-cutting importance of critical minerals security for national security, economic competitiveness, and the energy transition. Minerals are the bedrock of military technologies, semiconductors, electric vehicles (EVs), and clean energy technologies. The United States' overreliance on China has already brought dire consequences, as evidenced by the full ban on germanium, gallium, and antimony exports to the United States, alongside strict restrictions on graphite in December 2024. Addressing these challenges is not merely an economic necessity, but a national security imperative.

Strengthening U.S. critical minerals supply chains demands a comprehensive strategy focused on domestic resource development, advanced processing and recycling technologies, international partnerships, and sustainable practices. This volume explores these strategies and provides concrete recommendations for multiple dimensions of minerals security to ensure resilience and self-reliance in critical minerals supply chains.

This volume has three primary objectives with regards to critical minerals and the future of the U.S. economy. First, it shows dependence of the twenty-first-century U.S. economy on a wide range of minerals for industries such as defense, semiconductors, and clean energy—and identifies where key minerals vulnerabilities exist within these supply chains. Second, it evaluates the policies and legislative initiatives undertaken during the Biden administration, identifying both strengths and weaknesses, before providing recommendations to enhance these efforts. And finally, it puts forward a comprehensive set of recommendations in key areas for the new administration, including domestic permitting, midstream processing, international engagement, and responsible mining.

This volume is intended to be a resource for policymakers in the new U.S. administration and Congress, written by experts with deep expertise working with industry and government. This book shows that critical minerals have never been more important for the U.S. economy than at this critical juncture. And yet significant vulnerabilities continue to exist in the supply chain, despite the ongoing efforts by government and businesses to address them.

Three prevailing themes emerge from the rich and diverse analysis contained in these chapters. First, critical minerals are ever-more pervasive in the modern U.S. economy and crosscut all facets of defense, economic security, and energy security. Second, although the U.S. government has taken credible steps and has invested hundreds of billions of dollars in addressing vulnerabilities in the critical minerals supply chain, it remains dependent on foreign adversaries—primarily China, but also Russia for minerals such as palladium. Lastly, given the dispersion of mineral resources globally, the United States will not be able to achieve full independence in the supply chain on its own. Addressing domestic permitting bottlenecks will need to go hand-in-hand with working with allies in developed countries and the Global South to break China's stranglehold on the global critical minerals market.

A fundamentally important element of any new approach to critical minerals policy must be undertaken in a strategic and integrated fashion. The interconnected policy recommendations put forward by the authors in this volume address the challenges of critical minerals dependency by promoting economic security, sustainability, and innovation. The United States urgently requires a comprehensive critical minerals strategy. By developing such an integrated approach, the United States will strengthen the foundation for industries essential to the future, including defense, semiconductors, EVs, and renewable energy.

Ten key recommendations emerged in this volume:

1. **Develop a comprehensive incentives package for mineral production and processing.**

   The administration should create a comprehensive incentives package focusing on minerals production and processing for vital industries, including defense, semiconductors, EVs, and energy. This strategy must go further than the CHIPS and Science Act, which was developed to bolster domestic semiconductor manufacturing but failed to include any support for securing the minerals needed to produce the semiconductors. Any new strategy must also be more aggressive than the Inflation Reduction Act (IRA), which created useful incentives to invest in critical minerals production and processing but was self-limiting, given that it only applied to the

United States and countries with which it had a free trade agreement. The incentives must support domestic and international mineral production. The lack of sufficient U.S. mining supply chain investment incentives abroad has allowed China to gain control of a substantial share of resources in Latin America and Africa and export them back to China for processing. If the United States is to meaningfully loosen China's chokehold on mineral production, it must implement incentives that reach both producers at home and investors abroad.

Minerals projects in the United States are struggling to stay competitive and remain operational amid low commodity prices and intense competition from China. Chinese mineral operations tend to use cheaper energy sources and labor to keep costs low. Additionally, projects are often heavily state subsidized, with the government serving as a buyer of last resort to manipulate markets and keep prices low. As a result, Western producers are unable to compete and in some cases are producing at a net loss. As a result, Western investors are less eager and more hesitant to invest in minerals projects in the United States and in partner nations. Incentives help these strategic projects to remain competitive and profitable by providing capital support upfront that offsets the costs and risks of mineral investments. Comprehensive incentives packages that target minerals beyond those used in EV batteries are a necessity to ensure Western private industry makes the investments now that will yield the required minerals for strategic industries in the near future.

There has thus far been insufficient investment in the critical minerals needed for advanced semiconductor technologies, particularly gallium and germanium. Implementing tax credits for midstream production of minerals domestically and in key allied countries can help to ensure these complex projects get off the ground. For EVs, the 30D provision of the IRA should be modified to expand benefits beyond solely free trade agreement (FTA) countries. Few mineral-rich countries currently hold an FTA-equivalent agreement with the United States, and are therefore excluded from IRA benefits. This is a critical step to incentivize mineral-rich nations to secure offtake agreements with Western countries instead of China.

FEOC rules should be tightened for incentives across industries. The current FEOC threshold of 25 percent ownership is too high and means projects with significant Chinese ownership are receiving benefits subsidized by U.S. taxpayers. However, given the limited number of projects outside of China for some key minerals, eliminating all tax benefits immediately could undermine market confidence and lead to supply chain disruptions. Instead, FEOC rules should be gradually phased down to 0 percent so that only projects that align with U.S. interests receive benefits, bolstering the production of minerals that support the semiconductor, EV, energy, and defense industries.

2. **Invest in innovation.**

The U.S. economy continues to be among the most innovative in the world, and the power of this innovation must be harnessed to drive technological progress, improve competitiveness, and reduce the country's critical minerals vulnerabilities. Investments in building recycling capabilities also reduce the quantity of primary minerals that need to be mined in the future, reducing the risk of future shortages. Innovation will be particularly critical for the semiconductor and defense industries, which are the mainstay of the country's national and economic security.

- **Semiconductors:** After initially imposing export restrictions on gallium and germanium in 2024, China escalated to a full export ban of these two commodities to the United States in December 2024. The United States needs to rapidly build the technological expertise for refining gallium and germanium. Achieving the required purity levels of over 99.99 percent for the semiconductor industry necessitates specialized technology, infrastructure, and knowledge, which are currently absent. Currently, the United States has just one company refining high-purity gallium and only one facility for germanium. Establishing a research and development (R&D) lab could foster innovation, enhance processing capacity, reduce the industry's environmental footprint, and reduce production costs. Targeted funding through

CHIPS Act initiatives, such as the National Institute of Standards and Technology's R&D program and the Microelectronics Commons, would support advanced materials research, facilitate commercialization, and drive the next generation of semiconductor manufacturing.

- **EVs:** Given the importance of sustaining the economic competitiveness of the domestic automotive industry, the new administration should invest in R&D into cost-efficient, resource-abundant battery technologies and recycling technologies to strengthen the circular economy. Innovative battery technologies and the development of substitutes for scarce materials are key to alleviating resource pressures. Ongoing research into alternative battery technologies that reduce or eliminate reliance on critical minerals like cobalt must be supported by grants and collaborations with universities and startups.

Just as the Department of Energy funds laboratories for critical minerals in EVs and clean energy, the Department of Commerce should support similar initiatives for semiconductor minerals like gallium and germanium to scale refining capabilities and technologies. Likewise, the Department of Defense should similarly adopt efforts for minerals like tungsten and antimony.

3. **Strengthen stockpiling.**

The United States should establish a Critical Minerals Reserve to stabilize supply chains, mitigate geopolitical and market risks, and reduce dependence on foreign-controlled resources, using authorized market makers backed by federal loans to procure minerals from approved jurisdictions and support domestic and allied industries. The National Defense Stockpile (NDS) was created under the 1939 Strategic and Critical Materials Stock Piling Act to provide for the acquisition of critical materials to meet defense industry needs in cases of national emergency.[575] In the event of a natural disaster, regional conflict, or large-scale war, government-owned physical reserves of minerals can provide both the U.S. military and commercial industry with minerals needed for national defense.

Strategic stockpiles are increasingly put forth as a potential policy tool to not only provide materials in a national security emergency but also as a mechanism to help stabilize commodity markets. A government-run minerals stockpile can serve as a buyer of last resort for private industry. When global prices are too low for U.S. producers to compete due to subsidized Chinese production, the government could purchase minerals for the stockpile to drive demand and raise prices.

To strengthen mineral stockpiles, Congress should boost discrete program appropriations for the NDS Transaction Fund. From 1968 to 2022, Congress appropriated no new budget authority for minerals stockpiling. Over the past few decades, the minerals stockpile was depleted significantly, as the Department of Defense determined that over 99 percent of the stockpile was excess to the department's needs and Congress soon authorized its disposal.[576] As of 2023, the stockpile's value was estimated at just 1.2 percent of its 1962 value when adjusted for inflation.[577] Overall, the twenty-first century has seen minimal policy action around mineral stockpiling. Consequently, the NDS is significantly smaller and less powerful as a market mechanism than China's minerals stockpile.

4. **Adopt a coordinated government approach.**

The United States will need to strengthen both its mission clarity and coordination. At present, the U.S. government has yet to agree on a single critical minerals list, lacks a coordinating agency, and has incomplete incentives. The U.S. government should have a single point of accountability to oversee and coordinate the administration's multiple lines of effort. There are currently 15 government departments and agencies working on critical minerals. While the increased interest is a positive development, the lack of clear oversight and management of such a complex set of issues can lead to inefficiencies or agencies working at cross-purposes, which may frustrate the mission.

5. **Support the implementation of responsible mining practices.**

One of the consistent grievances with some Chinese firms is their poor environmental, social, and governance practices. This includes heavy deforestation and land degradation, low wages, failure to employ local workers, and use of bribery.[578] Adopting responsible mining practices can differentiate the United States from China and make it a more attractive investment partner.

Environmentally and socially responsible standards are vital in mining to address the sector's significant environmental, social, and governance challenges. They promote sustainable practices by reducing pollution, protecting biodiversity, and ensuring efficient resource use. Such standards also emphasize fair community engagement, worker safety, and ethical governance, helping to build trust and manage risks like corruption and reputational damage. They also provide helpful frameworks to be applied to new emerging frontiers for mineral extraction, like deep sea mining.

- With growing investor and consumer demand for responsibly sourced materials, adherence to environmental and social principles improves social license to operate, regulatory compliance, access to green financing, economic feasibility, and long-term profitability for minerals projects on land and under the sea.
- Voluntary standards should be consolidated into one global system with independent oversight to streamline compliance and increase the effectiveness of industry-imposed standards that are currently inconsistent and cumbersome. This could be achieved by creating a national panel on mining and incorporating standards into trade agreements.
- The federal government should fund research and pilot projects for deep sea mineral processing and work toward full-scale projects with support from programs like the Inflation Reduction Act's 48C credit, the Defense Production Act, and the Loan Programs Office.

6. **Bolster financing for minerals supply chain development.**

The U.S. International Development Finance Corporation (DFC), Defense Production Act (DPA), and Inflation Reduction Act (IRA) provide significant grant funding, loan guarantees, and tax credits that offer vital sources of capital for minerals projects. Without these financing mechanisms, few Western projects would be profitable, and few companies would be willing to invest in risky but strategic projects abroad. With government financing support, companies like Lynas Rare Earths, Syrah Resources, Albemarle Corporation, and TechMet are initiating and expanding mining and processing capacity both in the United States and in foreign partner countries. But current financing mechanisms are still not enough to compete with foreign adversaries and secure mineral resources around the world.

The limitations of the DFC, DPA, and IRA programs leave many key mining jurisdictions outside of financing mechanisms.

- The DFC can be amended by Congress to expand its impact by allowing financing in high-income countries. The DFC's role in fostering minerals security is still fairly new, but a number of DFC-backed investments in Brazilian nickel, South African copper, and Angolan rare earths are already making an impact by helping to mitigate risks for strategic projects in mineral rich but underdeveloped nations. The DFC can continue to foster government-to-government cooperation and grow its impact by expanding its authorities and the number of eligible nations.
- The DPA spending authority should be increased while making greater use of purchasing commitments. DPA funds can now be used for projects in not only the United States and Canada but also Australia and the United Kingdom. More eligible jurisdictions mean more strategic projects can be funded to close supply chain gaps for defense minerals. More DOD purchase commitments can sustain the capabilities that have received investments and provide industry with the demand signal and business case to make their own additional investments.
- Price floors are a key financing support mechanism that gives projects assurance they will remain profitable even in the face of volatile and falling commodity prices. With this price support, industry is more likely to make strategic investments, knowing their products will sell at a fixed price without being subject to uncertain market conditions

vulnerable to Chinese manipulation. Price floors hedge risks and keep operational key minerals projects that serve U.S. interests.

Together, these policy adjustments will augment current financing programs and allow financial support to reach mineral-rich jurisdictions that were previously left out of these opportunities.

7. **Develop the critical minerals workforce.**

The Society for Mining, Metallurgy, and Exploration has estimated that more than half of the current workforce will be retired and replaced by 2029 (roughly 221,000 workers).[579] A skilled workforce is crucial for sustaining industries reliant on critical minerals. Throughout this volume, authors have recognized the need for investments into education, training, and lifelong career learning. Workforce development programs focusing on technical training in mining, materials science, processing and refining, defense applications, and advanced manufacturing will prepare the American labor force to support resilient domestic supply chains and maintain technological leadership.

- The new administration should work with Congress to establish a National Critical Minerals Workforce Initiative, leveraging federal programs like WIOA and tax incentives. This would help address workforce challenges by supporting education in key fields, fostering community college certifications, and integrating regional training centers into the American Job Centers network.
- Increasing investments in workforce development, leveraging successful initiatives like the Industrial Base Analysis and Sustainment (IBAS) program's National Imperative for Industrial Skills (NIIS), would strengthen recruitment, training, and retention efforts in key sectors such as critical minerals through active partnerships and regionally focused activities.

8. **Streamline the domestic permitting process.**

If the United States is to effectively boost domestic production of critical minerals, a number of serious challenges in the domestic permitting process must be overcome. A streamlined permitting process reduces delays, lowers costs for mining companies, and encourages private sector investment in domestic projects, reducing dependence on foreign-controlled supply chains. It also enhances the efficiency and predictability of regulatory reviews, enabling companies to plan and execute projects more effectively while maintaining environmental and community safeguards. By balancing environmental, social, and governance concerns with expedited approvals, a reformed permitting process supports a resilient and competitive critical minerals industry, bolsters economic growth, and ensures that the United States can secure a stable supply of these essential materials for its energy transition and technological needs. The authors in this book have suggested the following concrete initiatives:

- Congress should increase funding for agencies to hire more mineral experts, enabling timely National Environmental Policy Act (NEPA) reviews by improving early engagement with mine project applicants and reducing delays caused by insufficiently detailed mine plans. Congress should also allocate additional funding to defray mine applicants' permitting costs, including support for hiring approved, high-quality contractors to streamline the NEPA process and reduce compliance burdens.
- The new administration should direct the Permitting Council to use its voting authority to include more mining projects under FAST-41, bypassing the $200 million investment threshold for projects extracting critical energy transition minerals.
- The new administration's Council on Environmental Quality (CEQ) should issue a rule establishing clear thresholds for "intensity" factors in the NEPA process, enabling agencies to determine when an environmental impact assessment (EIS) is sufficient and avoiding unnecessary environmental impact statement preparation. Increasing the use of programmatic environmental reviews and categorical exclusions would streamline permitting for critical minerals projects by pre-evaluating broad categories of activities, reducing

redundancy in individual project assessments.

- Congress could pass legislation categorically exempting mines producing primary energy transition minerals from NEPA's environmental assessment and EIS requirements, expediting permitting timelines while specifying covered minerals from the Department of Energy's critical energy materials list.

9. Pursue strategic international engagement.

The United States will need to take a page out of China's playbook if it is to meaningfully compete. Over the past 30 years, China has leveraged its foreign policy to establish a dominant position in global critical minerals markets. Although it produces only 10 percent of the world's lithium, cobalt, nickel, and copper, China has made strategic investments globally and imports enough to process 65 to 90 percent of the global supply of these metals at home.[580] The United States will need to collaborate with historical allies—like Australia, Canada, and the European Union—in addition to Global South countries. This will require a judicious balance of sticks and carrots.

- The administration should prioritize stable tariff policies and minimize tariffs with allied nations to maintain investor confidence and minimize potential supply chain disruptions. It should balance tariffs and restrictions on Chinese products to address market manipulation, unfair practices, carbon intensity, and human rights without applying a blanket exclusion on Chinese firms, given that non-Chinese supply chains for a number of commodities are currently underdeveloped or undeveloped.

- The U.S. government can align its diplomacy efforts with its financing mechanisms—for example, priority financial support can be given to partners and forum members in the Minerals Security Partnership (MSP). It can also reform existing financing instruments to enable greater flexibility to fund supply chain diversification projects. For example, the 2025 reauthorization of the DFC provides an opportunity for the institution to go beyond its current mandate and finance mining projects in high-income allied countries like Canada and Chile.

- The United States should utilize technical assistance. This includes supporting the geological mapping of resource-rich countries, which can de-risk investments for exploration companies seeking to invest in emerging economies. This is vital because over 99 percent of exploration projects are unsuccessful.[581] This effort also includes investing in infrastructure. The mining industry is highly energy and water intensive and requires rail and port infrastructure to be exported. China has gained a significant upper hand through infrastructure investment in resource-rich developing countries. During the first 10 years of China's Belt and Road Initiative (BRI), it invested $1 trillion.[582] These investments have allowed China to secure raw materials for domestic processing.

10. Strengthen the defense industrial base.

China is rapidly expanding its investments in munitions and acquiring advanced weapons systems and equipment at a pace five to six times faster than the United States.[583] While China operates with a wartime mindset to enhance its military readiness, the United States maintains a peacetime approach. Even before the implementation of new restrictions, the U.S. defense industrial base struggled with insufficient capacity and limited surge capabilities to meet production demands for defense technologies. Restrictions on critical mineral inputs will only exacerbate this gap, enabling China to further outpace the United States in developing these capabilities.

- The Joint Staff should develop a war-planning scenario aligned with DOD policy and National Defense Strategy objectives. This updated scenario would enable the generation of more realistic estimates for defense-related critical minerals requirements, ensuring adequate preparedness for future conflicts.

- The DOD should stabilize funding for critical minerals in the base budget to ensure predictability for industry and investment planning, moving beyond reliance on one-off supplemental appropriations by utilizing mechanisms like discrete program increases or functional transfers within industrial mobilization programs.

- The DOD should streamline critical minerals sourcing rules by conducting an acquisition reform study to identify and consolidate overlapping regulations, simplifying exception structures. It should also develop a legislative proposal to reduce compliance burdens while supporting sub-tier supplier validation and military specification development. Furthermore, delegating authority for approving Presidential Determinations to the secretary of defense would significantly shorten the timeline for industrial base investments while ensuring informed decisions on critical minerals shortfalls with input from relevant departments and agencies.

ABOUT THE EDITORS

Dr. Gracelin Baskaran is the director of the Critical Minerals Security Program at the Center for Strategic and International Studies (CSIS) and a bye-fellow in economics at the University of Cambridge. She is a mining economist, and her primary expertise is in critical minerals and trade. She began her career in South Africa's platinum belt. She later spent five years at the World Bank in South Africa, where she coauthored the book *Africa's Resource Future: Harnessing Natural Resources for Economic Transformation during the Low-Carbon Transition*. She was a Fulbright Scholar and has also held positions at the University of Cambridge, University of London, and University of Cape Town. Dr. Baskaran is now writing a book on an international strategy for critical minerals engagement for the United States. She has been cited as an expert in the *The Economist, Financial Times, Wall Street Journal, Washington Post,* Bloomberg, Foreign Policy, and the U.S. International Trade Commission, among others. She has also published over 150 op-eds and columns, policy reports, and white papers and published research in the *Mineral Economics* journal. She has testified before the U.S. Congress and is a regular speaker at leading international mining and energy conferences, universities, and policymaker forums. Dr. Baskaran holds a doctorate from the University of Cambridge.

Dr. Duncan Wood is president and CEO of the Pacific Council of International Policy. He was previously the vice president for srategy and new initiatives at the Wilson Center, where he worked on supply chain policy, critical minerals, electric vehicles, and geopolitics of the energy transition. From 2013 to 2020, he was director of the Wilson Center's Mexico Institute and served as a senior advisor to the institute. He regularly gives testimony to the U.S. Congress, is a widely quoted media source, and has published extensively on a wide range of global issues. He is the author or editor of 12 books and more than 30 chapters and articles. He is a board member of Transparency International, Signos Vitales (a Mexican public policy research organization) and Foreign Affairs Latinoamerica, and is an editorial advisor to El Universal newspaper. Over the past decade he has served as co-chair of the World Economic Forum (WEF)'s Global Future Council on Transparency and Anti-Corruption and has worked closely with the WEF on energy policy. From 1996 to 2012, he was a professor and the director of the International Relations Program at the Instituto Tecnologico Autonomo de Mexico (ITAM) in Mexico City. Over the course of his 30-year career, he has been a Fulbright Scholar and a visiting fellow at the Center for Strategic and International Studies (CSIS). He holds a doctorate from Queen's University.

ABOUT THE CONTRIBUTORS

Dr. Morgan Bazilian is the director of the Payne Institute and a professor of public policy at the Colorado School of Mines. He is a senior associate (non-resident) of the Energy Security and Climate Change Program at CSIS.

Rohitesh Dhawan is the president and CEO of the International Council on Mining and Metals (ICMM) and a senior associate (non-resident) of the Critical Minerals Security Program at CSIS.

Frank Fannon is the managing director of Fannon Global Advisors and a senior adviser (non-resident) to the Critical Minerals Security Program at CSIS. Formerly, he served as assistant secretary of state for energy resources at the U.S. Department of State (2018–2021).

Alexandra Helfgott is an MPP candidate at the Yale Jackson School of Global Affairs.

Adam Johnson is CEO of Principal Mineral and managing partner at Metis Endeavor. Formerly, he served as senior vice president of corporate development and strategy at MP Materials.

Dr. Joseph Majkut is the director of the Energy Security and Climate Change Program at CSIS.

Christine Michienzi is the founder and CEO of MMR Defense Solutions LLC and a senior associate (non-resident) of the Defense and Security Department at CSIS. Formerly, she served as chief technology officer at the Office of the Secretary of Defense (2020–2023).

Jane Nakano is a senior fellow with the Energy Security and Climate Change Program at CSIS.

Meredith Schwartz is a research associate with the Critical Minerals Security Program at CSIS.

Dr. Seaver Wang is the co-director of the Climate & Energy Program at the Breakthrough Institute.

Kellee Wicker is the director of the Science and Technology Innovation Program at the Wilson Center.

Gregory Wischer is the founder and principal of Dei Gratia Minerals.

Matthew Zolnowski is the president of Greyfriars LLC. Formerly, he served as portfolio manager for DPA Investments and special advisor for strategic materials at the U.S. Department of Defense.

ENDNOTES

1 Linda R. Rowan, *Critical Mineral Resources: The U.S. Geological Survey (USGS) Role in Research and Analysis*, CRS Report No. R48005 (Washington, DC: Congressional Research Service, April 2024), https://crsreports.congress.gov/product/pdf/R/R48005.

2 Gracelin Baskaran and Meredith Schwartz, "China Imposes Its Most Stringent Critical Minerals Export Restrictions Yet Amidst Escalating U.S.-China Tech War," CSIS, *Critical Questions*, December 4, 2024, https://www.csis.org/analysis/china-imposes-its-most-stringent-critical-minerals-export-restrictions-yet-amidst.

3 The White House, *Building Resilient Supply Chains, Revitalizing American Manufacturing, and Fostering Broad-Based Growth*, (Washington, DC: The White House, 2021), https://www.whitehouse.gov/wp-content/uploads/2021/06/100-day-supply-chain-review-report.pdf.

4 Strategic and Critical Minerals Stockpiling Act, Pub. L. 118-31, June 7, 1939, https://www.govinfo.gov/content/pkg/COMPS-674/pdf/COMPS-674.pdf.

5 "Letter on Appropriations for Strategic War Materials," American Presidency Project, accessed December 13, 2024, https://www.presidency.ucsb.edu/documents/letter-appropriations-for-strategic-war-materials.

6 "Events Affecting the U.S. Nonfuel Minerals Industry 1900-2000," Upper South Platte Partnership, March 2015, https://uppersouthplatte.org/learn/wp-content/uploads/2015/03/timeline.pdf.

7 Alexandra Neenan and Luke Nicastro, *The Defense Production Act of 1950: History, Authorities, and Considerations for Congress*, CRS Report No. R43767 (Washington, DC: Congressional Research Service, October 2023), https://crsreports.congress.gov/product/pdf/R/R43767.

8 "Events Affecting the U.S. Nonfuel Minerals Industry 1900-2000," Upper South Platte Partnership; and "Catalytic Converter Ensuring 90% Carbon Emissions' Reduction," WIPO, accessed December 13, 2024, https://www.wipo.int/web/ip-advantage/w/stories/catalytic-converter-ensuring-90-carbon-emissions-reduction.

9 S&P Global Markets in Motion, Copper On The Rise (New York: S&P Global, 2022), https://www.spglobal.com/en/research-insights/special-reports/copper-on-the-rise.

10 Joseph Hoppe and Christian Moess Laursen, "AI Siphons Copper Supplies Needed for Green Transition," *Wall Street Journal*, June 6, 2024, https://www.wsj.com/articles/ai-siphons-copper-supplies-needed-for-green-transition-8fef79e6.

11 Creamer Media's Mining Weekly, "Copper Hits Record above $11 000 on Bets That Shortage Looms," Bloomberg, May 20, 2024, https://www.miningweekly.com/article/copper-surges-to-record-as-investors-bet-on-looming-shortage-2024-05-20.

12 "Member Agencies," Critical Minerals Subcommittee, accessed December 13, 2024, https://www.criticalminerals.gov/pages/member-agencies.

13 "The Need to De-Risk Gallium Material Supply Chains," Microwave Journal, accessed December 13, 2024, https://www.microwavejournal.com/articles/41212-the-need-to-de-risk-gallium-material-supply-chains.

14 Seth Jones, "The U.S. Defense Industrial Base Is Not Prepared for a Possible Conflict with China," CSIS, February 2023, https://features.csis.org/preparing-the-US-industrial-base-to-deter-conflict-with-China/.

15 "Events Affecting the U.S. Nonfuel Minerals Industry 1900-2000," Upper South Platte Partnership.

16 Chris Busch, "Inflation Reduction Act Two Years Later: Auto Manufacturing Jobs At 34-Year Peak," Forbes, August 21, 2024, https://www.forbes.com/sites/energyinnovation/2024/08/21/inflation-reduction-act-two-years-later-auto-manufacturing-jobs-at-34-year-peak/.

17 "Clean Energy Investment in Every State," Climate Central, accessed December 13, 2024, https://www.climatecentral.org/climate-matters/clean-energy-investment-in-every-state.

18 "Job Creation & Economic Growth," U.S. Department of Energy, accessed December 13, 2024, https://www.energy.gov/eere/job-creation-economic-growth.

19 "Clean Energy Investment in Every State," Climate Central.

20 Gracelin Baskaran, "Seven Recommendations for the New Administration and Congress: Building U.S. Critical Minerals Security," CSIS, *Commentary*, November 14, 2024, https://www.csis.org/analysis/seven-recommendations-new-administration-and-congress-building-us-critical-minerals; and Wayne Chancellor, Wayne and D. Scott Anderson, "Understanding Mining Permit Reform," National Law Review, April 2, 2024, https://natlawreview.com/article/permitting-reform-united-states.

21 Shiona McCallum, "What Are Semiconductors and How Are They Used?," BBC, August 3, 2023, https://www.bbc.com/news/technology-66394406.

22 "What Is a Semiconductor and How Is It Used?," Ideal Power, January 28, 2022, https://www.idealpower.com/what-is-a-semiconductor/.

23 "Semiconductor Materials," Institute of Electrical and Electronics Engineers, accessed October 3, 2024, https://irds.ieee.org/topics/semiconductor-materials.

24 "Fundamentals: Doping: N- and p-Semiconductors," Halbleiter, accessed October 3, 2024, https://www.halbleiter.org/en/fundamentals/doping/.

25 "In Which Industries or Applications Is Palladium Electroplating Most Prevalent, and Why?," ProPlate, May 8, 2024, https://www.proplate.com/in-which-industries-or-applications-is-palladium-electroplating-most-prevalent-and-why/; "Ti for Semiconductor Applications," Umicore Group, January 25, 2018, https://mds.umicore.com/storage/eom/tfp/ti.pdf;

and Katherine Bourzac, "Cobalt Could Untangle Chips' Wiring Problems," IEEE Spectrum, January 31, 2018, https://spectrum.ieee.org/cobalt-could-untangle-chips-wiring-problems.

26 Christopher Cytera, "Gallium, Germanium, and China – The Minerals Inflaming the Global Chip War," Center for European Policy Analysis, August 8, 2023, https://cepa.org/article/china-gallium-and-germanium-the-minerals-inflaming-the-global-chip-war/; and U. S. Geological Survey, *Mineral Commodity Summaries 2024* (Reston, VA: U.S. Geological Survey, 2024), https://doi.org/10.3133/mcs2024.

27 Ibid., 23, 36, 74, 161, 186.

28 "Price of China's Strategic Germanium Hits Record High on Possible State Buying," Reuters, July 18, 2024, https://www.reuters.com/markets/commodities/price-chinas-strategic-germanium-hits-record-high-possible-state-buying-2024-07-18/.

29 Gracelin Baskaran and Meredith Schwartz, "China Imposes Its Most Stringent Critical Minerals Export Restrictions Yet Amidst Escalating U.S.-China Tech War," CSIS, *Critical Questions*, December 4, 2024, https://www.csis.org/analysis/china-imposes-its-most-stringent-critical-minerals-export-restrictions-yet-amidst.

30 Gregory Arcuri and Sujai Shivakumar, "Moore's Law and Its Practical Implications," CSIS, *Critical Questions*, October 18, 2022, https://www.csis.org/analysis/moores-law-and-its-practical-implications.

31 Marissa Lee, "Silicon Is Reaching Its Limit. What's Next?," *Business Times* (Singapore), April 9, 2020, https://gansystems.com/newsroom/article-silicon-is-reaching-its-limit-whats-next/.

32 Peide D. Ye, "Germanium Can Take Transistors Where Silicon Can't," IEEE Spectrum, November 29, 2016, https://spectrum.ieee.org/germanium-can-take-transistors-where-silicon-cant; and Gérard Ghibaudo and Quentin Rafhay, "Electron and Hole Mobility in Semiconductor Devices," in John G. Webster ed., *Wiley Encyclopedia of Electrical and Electronics Engineering*, 1st ed. (New York: Wiley, 2000), 1-13, https://doi.org/10.1002/047134608X.W3148.pub2.

33 "5 Reasons Why You Should Make the Switch to Gallium Nitride (GaN) Over Silicon Devices," Kensington, November 12, 2021, https://www.kensington.com/news/docking-connectivity-blog/5-reasons-why-you-should-make-the-switch-to-gallium-nitride-gan-over-silicon-devices/.

34 "Gallium Nitride Semiconductor Devices Market Size, Share & Trends Analysis Report by Product, by Component, by Wafer Size, by End Use, by Region, and Segment Forecasts, 2023-2030," Grand View Research, June 14, 2024, https://www.grandviewresearch.com/industry-analysis/gan-gallium-nitride-semiconductor-devices-market.

35 Todd Helmenstine, "Abundance of Elements in Earth's Crust - Periodic Table and List," Science Notes, May 23, 2018, https://sciencenotes.org/abundance-of-elements-in-earths-crust-periodic-table-and-list/.

36 U.S. Geological Survey, *Mineral Commodities Summaries*, 75; and Madhav Patel and Athanasios K. Karamalidis, "Germanium: A Review of Its US Demand, Uses, Resources, Chemistry, and Separation Technologies," *Separation and Purification Technology* 275 (November 15, 2021): 118981, https://doi.org/10.1016/j.seppur.2021.118981.

37 "Germanium Prices," Strategic Metals Invest, accessed October 3, 2024, https://strategicmetalsinvest.com/germanium-prices/.

38 "Aluminum Ore," Observatory of Economic Complexity, accessed October 3, 2024, https://oec.world/en/profile/hs/aluminium-ore.

39 Matthew P. Funaiole, Brian Hart, and Aidan Powers-Riggs, "Mineral Monopoly: China's Control over Gallium Is a National Security Threat," CSIS, *Hidden Reach*, July 18, 2023, https://features.csis.org/hiddenreach/china-critical-mineral-gallium/.

40 U.S. Geological Survey, *Mineral Commodities Summaries*, 74-75.

41 Ibid., 43, 203.

42 Natalie Liu, "Tennessee Refinery Could Break Chinese Chokehold on Two Critical Minerals," Voice of America, August 15, 2023, https://www.voanews.com/a/tennessee-refinery-could-break-chinese-chokehold-on-two-critical-minerals/7226979.html.

43 Nyrstar, "Nyrstar's Middle Tennessee Mines Operations to Temporarily Pause Production," press release, October 31, 2023, https://www.nyrstar.com/resource-center/press-releases/nyrstars-middle-tennessee-mines-operations-to-temporarily-pause-production.

44 "Zinc Ore," Observatory of Economic Complexity, accessed October 3, 2024, https://oec.world/en/profile/hs/zinc-ore; "Aluminum Ore," Observatory of Economic Complexity, accessed October 3, 2024, https://oec.world/en/profile/hs/aluminium-ore.

45 "Cajamarquilla," Nexa Resources, accessed October 3, 2024, https://ri.nexaresources.com/operations_smelting/cajamarquilla/.

46 Wassen Mohammad, Adel Elomri, and Laoucine Kerbache, "The Global Semiconductor Chip Shortage: Causes, Implications, and Potential Remedies," *IFAC-PapersOnLine* 55, no. 10 (June 2022): 476-83, https://doi.org/10.1016/j.ifacol.2022.09.439.

47 Emilia David, "Where the CHIPS Act Money Has Gone," The Verge, June 7, 2024, https://www.theverge.com/24166234/chips-act-funding-semiconductor-companies.

48 The White House, "Fact Sheet: CHIPS and Science Act Will Lower Costs, Create Jobs, Strengthen Supply Chains, and Counter China," press release, August 9, 2022, https://www.whitehouse.gov/briefing-room/statements-releases/2022/08/09/fact-sheet-chips-and-science-act-will-lower-costs-create-jobs-strengthen-supply-chains-and-counter-china/.

49 David, "Where the CHIPS Act Money Has Gone."

50 Zeyi Yang, "China Just Fought Back in the Semiconductor Exports War. Here's What You Need to Know," *MIT Technology Review*, July 10, 2023, https://www.technologyreview.com/2023/07/10/1076025/china-export-control-semiconductor-material/.

51 Kirti Gupta, Chris Borges, and Andrea Leonard Palazzi, "Collateral Damage: The Domestic Impact of U.S. Semiconductor Export Controls," CSIS, *Commentary*, July 9, 2024, https://www.csis.org/analysis/collateral-damage-domestic-impact-us-semiconductor-export-controls.

52 Matthew Blackwood and Catherine DeFilippo, "Germanium and Gallium: U.S. Trade and Chinese Export Controls," U.S. International Trade Commission, March 29, 2024, https://www.usitc.gov/publications/332/executive_briefings/ebot_germanium_and_gallium.pdf.

53 Solomon Cefai and Nico Zhang, "China's Gallium Export Control Continues to Squeeze Market, despite Higher December Exports," Fastmarkets, February 6, 2024, https://www.fastmarkets.com/insights/chinas-gallium-export-control-continues-to-squeeze-market-despite-higher-december-exports/.

54 Matthew Funaiole, Brian Hart, and Aidan Powers-Riggs, "The Need to De-Risk Gallium Material Supply Chains," Microwave Journal, December 13, 2023, https://www.microwavejournal.com/articles/41212-the-need-to-de-risk-gallium-material-supply-chains.

55 "Price of China's Strategic Germanium Hits Record High on Possible State Buying," Reuters, July 18, 2024, https://www.reuters.com/markets/commodities/price-chinas-strategic-germanium-hits-record-high-possible-state-buying-2024-07-18/; and Harry Dempsey and Edward White, "China's Export Curbs on Semiconductor Materials Stoke Chip Output Fears," *Financial Times*, August 26, 2024, https://www.ft.com/content/9cd56880-4360-4e11-8c22-e810d3787e88.

56 Blackwood and DeFilippo, "Germanium and Gallium."

57 John Jacobs, Resilient Resource Reserve: *A Plan to Catalyze the American Critical Mineral Processing Industry* (Washington, DC: Bipartisan Policy Center, 2024), https://bipartisanpolicy.org/report/critical-minerals-reserve/.

58 "Gallium & Gallium Alloys," Indium Corporation, accessed October 3, 2024, https://www.indium.com/products/metals/gallium/.

59 U.S. Geological Survey, *Mineral Commodities Summaries*, 74, 80.

60 "Critical Materials Innovation Hub," Ames National Laboratory, accessed October 3, 2024, https://www.ameslab.gov/cmi; and Christina Procopiou, "National Labs Collaborate to Accelerate Critical Minerals and Materials Research," Lawrence Berkeley National Laboratory, April 2, 2024, https://eesa.lbl.gov/2024/04/02/national-labs-collaborate-to-accelerate-critical-minerals-and-materials-research/.

61 Kristian Burt, "Biden Administration Awards Intel up to $3 Billion under the CHIPS Act," CNBC, September 16, 2024, https://www.cnbc.com/2024/09/16/intel-awarded-up-to-3-billion-under-chips-act.html.

62 U.S. Geological Survey, *Mineral Commodities Summaries*, 74-75.

63 Cameron M. Keys, *Emergency Access to Strategic and Critical Materials: The National Defense Stockpile*, CRS Report No. R47833 (Washington, DC: Congressional Research Service, November 2023), https://sgp.fas.org/crs/natsec/R47833.pdf.

64 50 U.S.C. 98h-5.

65 Ibid. Based on 50 U.S.C. 98h-5(b).

66 For additional information, see Matthew D. Zolnowski, "Short War, Long War, and Industrial Policy," Wilson Center, November 30, 2023, https://www.wilsoncenter.org/article/short-war-long-war-and-industrial-policy.

67 Freedom of Information Act Electronic Reading Room, "NSC Meeting on National Defense Stockpile Goals," CIA.gov, May 13, 1985, https://www.cia.gov/readingroom/document/cia-rdp-87b00342r000100090001-7.

68 Keys, *Emergency Access*.

69 To safeguard limited distribution data, Table 1 summarizes only the *awarded* critical minerals efforts under the Defense Production Act Investments program.

70 Keys, *Emergency Access*.

71 Joint Chiefs of Staff, *Joint Publication 4-05, Joint Mobilization Planning* (Washington, DC: U.S. Department of Defense, 2018), https://irp.fas.org/doddir/dod/jp4_05.pdf.

72 Bureau of Industry and Security, "The Defense Priorities and Allocations Systems Training Course," U.S. Department of Commerce, November 2021, https://www.bis.doc.gov/index.php/documents/other-areas/strategic-industries-and-economic-security/1615-dpas-training-slides/file.

73 "Security of Supply," Office of the Assistant Secretary of Defense for Industrial Base Policy, accessed September 19, 2024, https://www.businessdefense.gov/security-of-supply.html.

74 10 U.S.C. 4863 and Defense Federal Acquisition Regulation Supplement (DFARS) 252.225-7009; and 10 U.S.C 4872 and DFARS 252.225-7052.

75 "Specialty Metals and National Security: Fiscal Year 2020," Specialty Steel Industry of North America, November 2019, https://www.ssina.com/wp-content/uploads/2019/11/Specialty-Metals-and-National-Security-FY2020.pdf.

76 Moshe Schwartz, "Sourcing Rare Earth Magnets Posing Challenges," *National Defense*, May 24, 2024, https://www.nationaldefensemagazine.org/articles/2024/5/24/viewpoint-sourcing-rare-earth-magnets-posing-challenges.

77 Materials covered by the rule include samarium-cobalt magnets, neodymium-iron-boron magnets, tungsten metal powders, tungsten heavy alloys, and tantalum metal and alloys. Of these, only neodymium-iron-boron magnets are covered by a

recycled content exception.

78 Office of the Secretary of Defense, "DoD Legislative Proposals - Fiscal Year 2023," U.S. Department of Defense, May 13, 2022, https://ogc.osd.mil/Portals/99/OLC%20FY%202023%20Proposals/13May2022Proposals.pdf.

79 Such as the appointment and convening of the new "Board of Directors" to oversee stockpile management (50 U.S.C. 98h-1).

80 Office of the Secretary of Defense, *Department of Defense Revolving Funds Justification/Overview, Fiscal Year (FY) 2024 Budget Estimates* (Washington, DC: Department of Defense, March 2023), https://comptroller.defense.gov/Portals/45/Documents/defbudget/fy2024/budget_justification/pdfs/06_Defense_Working_Capital_Fund/DoD_Revolving_Funds_J-Book_fy2024.pdf.

81 See 50 U.S.C. 98e and 50 U.S.C. 98h-6; 10 U.S.C. 4817; and 50 U.S.C. 4534. The latter includes those supporting the ammunition industrial base (Procurement of Ammunition, Army / BA 02 / BSA 10 / 0572EP1000).

82 Kevin M. Tansey et al., "Foreign Involvement and Materials Qualification in the Title III Program," Government Accountability Office, March 1994, https://www.gao.gov/assets/nsiad-94-74.pdf.

83 See Section 1080 of National Defense Authorization Act for Fiscal Year 2024, Pub. L. 118-31.

84 Examples of Canadian projects include critical mineral mining and processing (Fortune Minerals Ltd., Lomiko Metals Inc., Electra Battery Materials Corporation, Fireweed Metals Corp), batteries (Nano One Materials Corp.), and ammunition (General Dynamics - Ordnance and Tactical Systems Canada Inc.)

85 Supplemental appropriations include the Inflation Reduction Act ($250 million), the Ukraine Supplemental Appropriations Act ($500 million), and the CARES Act ($28 million), with $411 million awarded to rare earth projects from the base budget. See C. Todd Lopez, "DoD Looks to Establish 'Mine-to-Magnet' Supply Chain for Rare Earth Elements," Department of Defense, March 11, 2024, https://www.defense.gov/News/News-Stories/Article/Article/3700059/.

86 Office of the Secretary of Defense, "Fiscal Year 2025 Budget Estimates - Defense Production Act Purchases," U.S. Department of Defense, March 2024, https://comptroller.defense.gov/Portals/45/Documents/defbudget/FY2025/budget_justification/pdfs/02_Procurement/PROC_DPAP_PB_2025.pdf.

87 See 16 U.S.C. 1536(j) and 42 U.S.C. 7412(i)(4).

88 See "Impact of National Environmental Policy Act on Certain Defense Industrial Base Projects" in *Report of the Committee on Armed Services on H.R. 8070*, House Report 118-529, May 31, 2024, U.S. House of Representatives, https://www.congress.gov/congressional-report/118th-congress/house-report/529/1.

89 South32 was admitted to the dashboard on May 5, 2023 (see "South32 Hermosa Critical Minerals Project," Permitting Dashboard, May 5, 2023, https://www.permits.performance.gov/permitting-project/fast-41-covered-projects/south32-hermosa-critical-minerals-project). Perpetua Resources' admittance date is not recorded (see "Stibnite Gold Project," Permitting Dashboard, accessed October 15, 2024, https://www.permits.performance.gov/permitting-project/other-projects/stibnite-gold-project).

90 Department of Defense, "Deputy Secretary of Defense Kathleen H. Hicks' Keynote Remarks: 'Strengthening the Transatlantic Arsenal of Democracy' [As Prepared]," NATO Summit Defense Industry Forum, U.S. Chamber of Commerce, Washington, D.C., July 9, 2024, https://www.defense.gov/News/Speeches/Speech/Article/3832532/deputy-secretary-of-defense-kathleen-h-hicks-keynote-remarks-strengthening-the/; and Gen. Christopher Mahoney, "Depth of magazine: Preparing the joint force for protracted conflict," *Military Times*, August 6, 2024, https://www.militarytimes.com/opinion/2024/08/06/depth-of-magazine-preparing-the-joint-force-for-protracted-conflict/.

91 Executive Office of the President, "Prioritizing and Allocating Health and Medical Resources to Respond to the Spread of COVID-19," *Federal Register* 85, no. 56 (March 23, 2020), https://www.federalregister.gov/documents/2020/03/23/2020-06161/prioritizing-and-allocating-health-and-medical-resources-to-respond-to-the-spread-of-covid-19.

92 Executive Office of the President, "Presidential Determination Pursuant to Section 303 of the Defense Production Act of 1950, as Amended," Government Printing Office, July 22, 2019, https://www.govinfo.gov/content/pkg/DCPD-201900500/pdf/DCPD-201900500.pdf.

93 For example, samarium-cobalt magnets under the "specialty metals clause" and the "sensitive materials rule" and the recycling exception of the "sensitive materials rule."

94 Office of the U.S. Trade Representative, *Report to Congress on the Operation of the United States-Mexico-Canada Agreement with Respect to Trade In Automotive Goods*, (Washington, DC: Office of the U.S. Trade Representative, 2022), https://ustr.gov/sites/default/files/2022%20USMCA%20Autos%20Report%20to%20Congress.pdf.

95 "North America in Motion: The Integration of the Auto Industry through Free Trade Agreements," México ¿Cómo Vamos? and the U.S. - Mexico Foundation, n.d., https://cdn.prod.website-files.com/64e7ab1532769a05d0365455/66fab82f47e9b74451608622_Auto%20Industry_TNAP_2024_ENG.pdf.

96 Office of the U.S. Trade Representative, *Report to Congress on the Operation of the United States-Mexico-Canada Agreement*.

97 "10 propuestas para el nuevo gobierno," Asociación Mexicana de la Industria Automotriz, n.d., https://www.amia.com.mx/.

98 Canadian Vehicle Manufacturers' Association, *2023 State of the Canadian Auto Industry*, (Ontario, Canada: Canadian Vehicle Manufacturers' Association, 2023), https://www.cvma.ca/wp-content/uploads/2023/12/State-of-the-Canadian-Automotive-Industry-2023.pdf

99 "North America in Data, Numbers, and Facts," The North American Project, México ¿Cómo Vamos? and the U.S.-Mexico

Foundation, 2024, https://cdn.prod.website-files.com/64e7a-b1532769a05d0365455/66fab861eaa656d5493e0571_Factsheet_TNAP%202024_ENG.pdf.

100 IEA (International Energy Agency), *Global EV Outlook 2024: Moving Towards Increased Affordability* (Paris: IEA, 2024), https://www.iea.org/reports/global-ev-outlook-2024/trends-in-electric-cars.

101 Ibid.

102 Neal E. Boudette and Coral Davenport, "G.M. Will Sell Only Zero-Emission Vehicles by 2035," *New York Times*, January 28, 2021, https://www.nytimes.com/2021/01/28/business/gm-zero-emission-vehicles.html.

103 "FACT SHEET: President Biden Announces Steps to Drive American Leadership Forward on Clean Cars and Trucks," The White House, August 5, 2021, https://www.whitehouse.gov/briefing-room/statements-releases/2021/08/05/fact-sheet-president-biden-announces-steps-to-drive-american-leadership-forward-on-clean-cars-and-trucks/.

104 "Credits for New Clean Vehicles Purchased in 2023 or After," Internal Revenue Service, last updated August 8, 2024, https://www.irs.gov/credits-deductions/credits-for-new-clean-vehicles-purchased-in-2023-or-after.

105 Ibid.

106 Ibid.

107 Peter Johnson, "Electric vehicle sales hit 7.9% of US market as Volvo, Mercedes, and Hyundai post sizable gains," electrek, October 12, 2023, https://electrek.co/2023/10/12/ev-sales-7-9-us-market-volvo-mercedes-hyundai-gain/

108 "Electric Vehicle Sales Mark Another Record in Q3, Thanks to Higher Incentives, More Choices," Cox Automotive, October 11, 2024, https://www.coxautoinc.com/market-insights/q3-2024-ev-sales/.

109 Michael Dwyer, "Electric Vehicles and Hybrids Surpass 16% of Total 2023 U.S. Light-Duty Vehicle Sales," U.S. Energy Information Administration, January 31, 2024, https://www.eia.gov/todayinenergy/detail.php?id=61344.

110 Michael Dwyer, "U.S. Share of Electric and Hybrid Vehicle Sales Decreased in the First Quarter of 2024," U.S. Energy Information Administration, May 14, 2024, https://www.eia.gov/todayinenergy/detail.php?id=62063.

111 "An association update from Chief Government Affairs and Policy Officer David Schwietert," Alliance for Automotive Innovation, October 15, 2024 https://www.autosinnovate.org/association-update/WM%20Tax%20Teams%20Letter%20-%20Auto%20Innovators.pdf.

112 Wilfried Eckl-Dorna, "European Car Sales Stagnate as EV Demand Plummets in Germany," Bloomberg, August 29, 2024, https://www.bloomberg.com/news/articles/2024-08-29/european-car-sales-stagnate-as-ev-demand-plummets-in-germany?embedded-checkout=true.

113 Ibid.

114 "Rare Earth Elements," Energy Systems Research Unit, University of Strathclyde, https://www.esru.strath.ac.uk/EandE/Web_sites/17-18/paradigmev/rare-earth-elements.html.

115 IEA, *Global Supply Chains of EV Batteries* (Paris: IEA, July 2022), https://iea.blob.core.windows.net/assets/961cf-c6c-6a8c-42bb-a3ef-57f3657b7aca/GlobalSupplyChainsofEVBatteries.pdf.

116 Haitham Al Ghais, "Critical Minerals: A Realistic Assessment," Organization of the Petroleum Exporting Countries, July 1, 2024, https://www.opec.org/opec_web/en/7346.htm.

117 IEA, Global EV Outlook 2024, (Paris: IEA, April 2024), https://www.iea.org/reports/global-ev-outlook-2024

118 IEA, *The Role of Critical Minerals in Clean Energy Technology*, (Paris: IEA, May 2021), https://www.iea.org/reports/the-role-of-critical-minerals-in-clean-energy-transitions.

119 Ibid.

120 Al Ghais, "Critical Minerals."

121 Saptakee S, "The Lithium Paradox: Price Plummet, Supply Surge, and Demand Dip - What's Happening Now?," CarbonCredits.com, September 16, 2024, https://carboncredits.com/the-lithium-paradox-price-plummet-supply-surge-and-demand-dip-whats-happening-now/.

122 Annie Lee, "Lithium Bust Spurs Ganfeng to Delay Project Capex as Losses Hit," Bloomberg, August 28, 2024, https://www.bloomberg.com/news/articles/2024-08-29/lithium-bust-spurs-ganfeng-to-delay-project-capex-as-losses-hit.

123 Ibid.

124 "Alternative Fuels Data Center," U.S. Department of Energy, accessed December 4, 2024, https://afdc.energy.gov/laws/search#/.

125 Matthew Zolnowski, *Ten Steps to Achieve Resilient Cobalt Supply Chains* (Washington, DC: Wahba Institute for Strategic Competition, Wilson Center, September 2024), https://www.wilsoncenter.org/publication/ten-steps-achieve-resilient-cobalt-supply-chains.

126 "An association update from Chief Government Affairs and Policy Officer David Schwietert," Alliance for Automotive Innovation; and Spencer Kimball and Gabriel Cortés, "How the Inflation Reduction Act Sparked a Manufacturing and Clean Energy Boom in the U.S.," CNBC, August 20, 2024, https://www.cnbc.com/2024/08/20/inflation-reduction-act-sparked-a-manufacturing-clean-energy-boom.html.

127 "An association update from Chief Government Affairs and Policy Officer David Schwietert," Alliance for Automotive Innovation.

128 James Bikales, "Why Biden's $7.5B Electric Vehicle Charger Push Is Probably Safe from Trump," *Politico*, November 20, 2024, https://www.politico.com/news/2024/11/20/biden-ev-charger-program-trump-proof-00190233.

129 IEA, *The Role of Critical Minerals*.

130 ReneSys Energy, "The Dominance of LFP in the Global Battery Market," ReneSys Energy, April 7, 2024, https://www.renesys.energy/news/the-dominance-of-lfp-in-the-global-battery-market.

131 Peter Campbell, Christian Davies, and Harry Dempsey, "Rival Battery Technologies Race to Dominate Electric Car Market," *Financial Times*, August 13, 2023, https://www.ft.com/content/67abc929-db1a-43c3-b03a-259a4316fd76.

132 Cosmo Sanderson, "Breakthrough for New Battery that Boasts Five Times the Power of Lithium-Ion," Recharge, October 19, 2023, https://www.rechargenews.com/energy-transition/breakthrough-for-new-battery-that-boasts-five-times-the-power-of-lithium-ion/2-1-1537524.

133 Alexandra Helfgott and Duncan Wood, *Competition, Climate, and Resilience: Securing the EV Supply Chain in America* (Washington, DC: Wilson Center, 2024), https://www.wilsoncenter.org/sites/default/files/media/uploads/documents/Competition%20Climate%20Report%20FINAL_0.pdf.

134 Costas Pitas, "Trump Vows New Canada, Mexico, China Tariffs that Threaten Global Trade," Reuters, November 26, 2024, https://www.reuters.com/world/us/trump-promises-25-tariff-products-mexico-canada-2024-11-25/.

135 International Energy Agency, "Clean Energy Investment Is Extending Its Lead Over Fossil Fuels, Boosted by Energy Security Strengths," press release, May 25, 2023, https://www.iea.org/news/clean-energy-investment-is-extending-its-lead-over-fossil-fuels-boosted-by-energy-security-strengths.

136 Energy Institute, *2024 Statistical Review of World Energy* (London: Energy Institute, 2024), 5, https://www.energyinst.org/_data/assets/pdf_file/0006/1542714/684_EI_Stat_Review_V16_DIGITAL.pdf.

137 Max Roser, "Why Did Renewables Become So Cheap So Fast?," Our World in Data, December 1, 2020, https://ourworldindata.org/cheap-renewables-growth.

138 "Frequently Asked Questions (FAQs): What Is U.S. Electricity Generation by Energy Source?," U.S. Energy Information Administration, accessed October 9, 2024, https://www.eia.gov/tools/faqs/faq.php?id=427&t=3.

139 Vikram Linga, "EIA Projects That Renewables Generation Will Supply 44% of U.S. Electricity by 2050," U.S. Energy Information Administration, March 18, 2022, https://www.eia.gov/todayinenergy/detail.php?id=51698; and Pieter Gagnon et al., *2023 Standard Scenarios Report: A U.S. Electricity Sector Outlook* (Golden, CO: National Renewable Energy Laboratory, 2024), vi, https://www.nrel.gov/docs/fy24osti/87724.pdf.

140 Aubryn Cooperman et al., *Renewable Energy Materials Properties Database: Summary* (Golden, CO: National Renewable Energy Laboratory, 2023), 11, 27, https://www.nrel.gov/docs/fy23osti/82830.pdf.

141 International Energy Agency, *The Role of Critical Minerals in Clean Energy Transitions* (Paris: International Energy Agency, 2021), 47–52, https://www.iea.org/reports/the-role-of-critical-minerals-in-clean-energy-transitions.

142 Seaver Wang et al., "Future Demand for Electricity Generation Materials under Different Climate Mitigation Scenarios," *Joule* 7, no. 2 (February 2023): 309–32, https://doi.org/10.1016/j.joule.2023.01.001.

143 Gracelin Baskaran, "What China's Ban on Rare Earths Processing Technology Exports Means," CSIS, *Commentary*, January 8, 2024, https://www.csis.org/analysis/what-chinas-ban-rare-earths-processing-technology-exports-means; Neil Ford, "US Solar Builders Brace for Higher Costs as Biden Hikes Tariffs," Reuters, May 23, 2024, https://www.reuters.com/business/energy/us-solar-builders-brace-higher-costs-biden-hikes-tariffs-2024-05-23/; and Jag Lamda, "Solar Panel Production Is Struggling to Stay Clear of Forced Labor," *PV Magazine*, March 28, 2024, https://pv-magazine-usa.com/2024/03/28/solar-panel-production-is-struggling-to-stay-clear-of-forced-labor/.

144 John Bistline et al., "Emissions and Energy Impacts of the Inflation Reduction Act," *Science* 380, no. 6652 (June 29, 2023): 1324–27, https://www.science.org/doi/10.1126/science.adg3781.

145 Cooperman et al., *Renewable Energy Materials*, 22.

146 Diana J. Bauer et al., *Critical Minerals Assessment* (Washington, DC: U.S. Department of Energy, 2023), xiii–xiv, https://www.energy.gov/sites/default/files/2023-07/doe-critical-material-assessment_07312023.pdf.

147 International Energy Agency, *The Role of Critical Minerals*, 8.

148 Mikayla Easley, "Special Report: U.S. Begins Forging Rare Earth Supply Chain," *National Defense*, February 10, 2023, https://www.nationaldefensemagazine.org/articles/2023/2/10/us-begins-forging-rare-earth-supply-chain.

149 U.S. Geological Survey, *Mineral Commodity Summaries 2024* (Reston, VA: U.S. Geological Survey, 2024), 144–45, https://pubs.usgs.gov/periodicals/mcs2024/mcs2024.pdf.

150 "What Are Rare Earth Elements?," MP Materials, accessed October 9, 2024, https://mpmaterials.com/what-we-do/.

151 MP Materials, "MP Materials Awarded $58.5 Million to Advance U.S. Rare Earth Magnet Manufacturing," press release, April 1, 2024, https://investors.mpmaterials.com/investor-news/news-details/2024/MP-Materials-Awarded-58.5-Million-to-Advance-U.S.-Rare-Earth-Magnet-Manufacturing/default.aspx.

152 Lynas Rare Earths, "Lynas Awarded US$120M Contract to Build Commercial HRE Facility," press release, June 14, 2022, https://wcsecure.weblink.com.au/pdf/LYC/02531153.pdf.

153 "Fact Sheet: Biden- Harris Administration Takes Further Action to Strengthen and Secure Critical Mineral Supply Chains," The White House, September 20, 2024, https://www.whitehouse.gov/briefing-room/statements-releases/2024/09/20/fact-sheet-biden-harris-administration-takes-further-action-to-strengthen-and-secure-critical-mineral-supply-chains/.

154 Katherine Antonio, "Solar and Wind to Lead Growth of U.S. Power Generation for the Next Two Years," U.S. Energy Infor-

mation Administration, January 16, 2024, https://www.eia.gov/todayinenergy/detail.php?id=61242.

155 Bistline et al., "Emissions and Energy Impacts."

156 Office of Energy Efficiency and Renewable Energy, *Solar Futures Study* (Washington, DC: U.S. Department of Energy, 2021), 33, https://www.energy.gov/sites/default/files/2021-09/Solar%20Futures%20Study.pdf.

157 Ibid.

158 Garvin Heath et al., *Environmental and Circular Economy Implications of Solar Energy in a Decarbonized U.S. Grid* (Golden, CO: National Renewable Energy Laboratory, 2022), 12, 103, https://www.nrel.gov/docs/fy22osti/80818.pdf.

159 Ibid., 103.

160 Office of Energy Efficiency and Renewable Energy, *Solar Futures Study*, 120–26.

161 Sylvia Leyva Martinez and Elissa Pierce, "Turn of the Tide? What the Entry of Chinese Polysilicon to the U.S. Means for the American Solar Supply Chain," Wood Mackenzie, December 18, 2023, https://www.woodmac.com/news/opinion/turn-of-the-tide-what-the-entry-of-chinese-polysilicon-to-the-us-means-for-the-american-solar-supply-chain/.

162 Nikos Tsafos, "Addressing Forced Labor Concerns in Polysilicon Produced in Xinjiang," CSIS, *Commentary*, June 7, 2021, https://www.csis.org/analysis/addressing-forced-labor-concerns-polysilicon-produced-xinjiang.

163 Haoyue Liang and Fengqi You, "Reshoring silicon photovoltaics manufacturing contributes to decarbonization and climate change mitigation," *Nature Communications* 14 (2023), https://www.nature.com/articles/s41467-023-36827-z.

164 Heath et al., *Environmental and Circular Economy*, 103.

165 "Solar Manufacturing Map," U.S. Department of Energy, accessed October 10, 2024, https://www.energy.gov/eere/solar/solar-manufacturing-map.

166 Hemlock Semiconductor, "Hemlock Semiconductor Breaks Ground on Next-Generation Finishing Facility," press release, October 21, 2022, https://www.hscpoly.com/hemlock-semiconductor-breaks-ground-on-nextgeneration-finishing-facility/; and Ivan Penn, "Key Solar Panel Ingredient Is Made in the U.S.A. Again," *New York Times*, April 25, 2024, https://www.nytimes.com/2024/04/25/business/energy-environment/climate-change-solar-panels-polysilicon-domestic.html.

167 Department of Energy, "Biden-Harris Administration Announces First Projects Receiving Clean Energy Manufacturing Investments in America's Industrial and Energy Communities," press release, April 19, 2024, https://www.energy.gov/articles/biden-harris-administration-announces-first-projects-receiving-clean-energy-manufacturing.

168 U.S. Department of Energy, "Federal Tax Credits for Solar Manufacturers," Office of Energy Efficiency and Renewable Energy, October 25, 2022, https://www.energy.gov/sites/default/files/2022-10/Federal-Tax-Credits-for-Solar-Manufacturers.pdf.

169 U.S. Department of Energy, "How LPO Can Support All Stages of the Critical Minerals Supply Chain," press release, April 30, 2024, https://www.energy.gov/lpo/articles/how-lpo-can-support-all-stages-critical-minerals-supply-chain.

170 Joseph Majkut et al., "Building Larger and More Diverse Supply Chains for Energy Minerals," CSIS, July 19, 2023, https://www.csis.org/analysis/building-larger-and-more-diverse-supply-chains-energy-minerals.

171 Jon Ossoff for Georgia, "Sen. Ossoff, Colleagues Press Biden Administration to Support U.S. Manufacturing of Strategic Solar Components," press release, February 29, 2024, https://www.ossoff.senate.gov/press-releases/sen-ossoff-colleagues-press-biden-administration-to-support-u-s-manufacturing-of-strategic-solar-components/.

172 "Fact Sheet: President Biden Takes Action to Protect American Workers and Businesses from China's Unfair Trade Practices," The White House, May 14, 2024, https://www.whitehouse.gov/briefing-room/statements-releases/2024/05/14/fact-sheet-president-biden-takes-action-to-protect-american-workers-and-businesses-from-chinas-unfair-trade-practices/.

173 Jon Ossoff for Georgia, "Sen. Ossoff Introduces Bipartisan Bill to Prevent Chinese Companies from Using American Solar Manufacturing Tax Credits," press release, August 1, 2024, https://www.ossoff.senate.gov/press-releases/sen-ossoff-introduces-bipartisan-bill-to-prevent-chinese-companies-from-using-american-solar-manufacturing-tax-credits/.

174 "FACT SHEET: President Biden to Catalyze Global Climate Action through the Major Economies Forum on Energy and Climate," The White House, April 20, 2023, https://www.whitehouse.gov/briefing-room/statements-releases/2023/04/20/fact-sheet-president-biden-to-catalyze-global-climate-action-through-the-major-economies-forum-on-energy-and-climate/.

175 Robinson Meyer, "Not Even a Single Republican Voted for the Climate Bill," *The Atlantic*, August 12, 2022, https://www.theatlantic.com/science/archive/2022/08/ira-climate-bill-house-vote-republicans/671133/.

176 Kristi E. Swartz, "GOP Gets 85% of the Benefit of Climate Law. Some Still Hate It.," Floodlight, August 14, 2024, https://floodlightnews.org/gop-gets-85-of-the-benefit/.

177 Daniel Yergin et al., "The IRA and the US's Mineral Supply Challenge," S&P Global Market Intelligence, S&P Global, August 15, 2023, https://www.spglobal.com/market-intelligence/en/news-insights/research/us-ira-and-critical-mineral-supply-challenge.

178 "Foreign Entity of Concern Interpretive Guidance," U.S. Department of Energy, accessed December 11, 2024, https://www.energy.gov/mesc/foreign-entity-concern-interpretive-guidance.

179 U.S. Department of Energy, "DOE Releases Final Interpretive Guidance on the Definition of Foreign Entity of Concern," press release, May 3, 2024, https://www.energy.gov/articles/

doe-releases-final-interpretive-guidance-definition-foreign-entity-concern.

180 Elaine Buckberg, "Clean Vehicle Tax Credit: The New Industrial Policy and Its Impact," Stanford Institute for Economic Policy Research, August 2023, https://siepr.stanford.edu/publications/policy-brief/clean-vehicle-tax-credit-new-industrial-policy-and-its-impact.

181 Jared Rosenholtz, "Only 10 Cars Will Get the $7,500 Electric Vehicle Tax Credit in 2024," CarBuzz, December 16, 2023, https://carbuzz.com/features/only-10-cars-will-get-the-7500-electric-vehicle-tax-credit-in-2024/.

182 "Role of Private Overseas Chinese Companies Key to US EV Tax Credits," Benchmark Source, December 8, 2023, https://source.benchmarkminerals.com/article/role-of-private-overseas-chinese-companies-key-to-us-ev-tax-credits.

183 Congresswoman Carol Miller, "Miller, Colleagues Introduce Bipartisan, Bicameral Congressional Review Act to Repeal Biden's Electric Vehicle Credit Rules," press release, May 16, 2024, http://miller.house.gov/media/press-releases/miller-colleagues-introduce-bipartisan-bicameral-congressional-review-act.

184 "Minerals Used in Electric Cars Compared to Conventional Cars," International Energy Agency, accessed December 9, 2024, https://www.iea.org/data-and-statistics/charts/minerals-used-in-electric-cars-compared-to-conventional-cars.

185 "Mineral Commodity Summaries: Graphite (Natural)," United States Geological Survey, January 2024, https://pubs.usgs.gov/periodicals/mcs2024/mcs2024-graphite.pdf; and Solomon Cefai, "CRMA 'Toothless' to Support European Graphite Producers; China, US Abandon Level Playing Field - Vianode CEO," Fastmarkets, June 13, 2024, https://www.fastmarkets.com/insights/crma-toothless-to-support-european-graphite-producers-china-us-abandon-level-playing-field vianode-ceo/.

186 Eri Silva, "North American Graphite Market to Disconnect from Chinese Prices," S&P Global Commodity Insights, S&P Global, August 15, 2023, https://www.spglobal.com/marketintelligence/en/news-insights/latest-news-headlines/north-american-graphite-market-to-disconnect-from-chinese-prices-76963809; and Graphite One, "Graphite One Awarded $37.5 Million Department of Defense Grant Under the Defense Production Act," press release, July 17, 2023, https://www.graphiteoneinc.com/graphite-one-awarded-37-5-million-department-of-defense-grant-under-the-defense-production-act/.

187 Jael Holzman, "Biden Exempts EV Metal from Climate Law Sourcing Rule," Axios, May 3, 2024, https://www.axios.com/2024/05/03/biden-graphite-ev-metal-climate-law-sourcing.

188 North American Graphite Alliance, "NAGA Urges Departments of Treasury and Energy to Classify Graphite as a Traceable Critical Mineral in Final Implementation of IRA's Section 30D Clean Vehicle Tax Credit," press release, April 2, 2024, https://graphitealliance.org/naga-urges-departments-of-treasury-and-energy-to-classify-graphite-as-a-traceable-critical-mineral-in-final-implementation-of-iras-section-30d-clean-vehicle-tax-credit/.

189 Internal Revenue Service, "Section 45X Advanced Manufacturing Production Credit," *Federal Register,* December 15, 2023, https://www.federalregister.gov/documents/2023/12/15/2023-27498/section-45x-advanced-manufacturing-production-credit.

190 Federal Register. "Section 45X Advanced Manufacturing Production Credit," December 15, 2023. https://www.federalregister.gov/documents/2023/12/15/2023-27498/section-45x-advanced-manufacturing-production-credit.

191 Internal Revenue Service, "Comment from National Mining Association," Regulations.Gov, February 7, 2024, https://www.regulations.gov/comment/IRS-2023-0063-0014.

192 Internal Revenue Service, "Advanced Manufacturing Production Credit," *Federal Register*, October 28, 2024, https://www.federalregister.gov/documents/2024/10/28/2024-24840/advanced-manufacturing-production-credit.

193 U.S. Department of the Treasury, "U.S. Department of the Treasury Releases Final Rules to Onshore Clean Energy Technologies, Strengthen Critical Minerals Supply Chains, and Expand U.S. Manufacturing Base as Part of Investing in America Agenda," press release, October 24, 2024, https://home.treasury.gov/news/press-releases/jy2671.

194 Wendolyn Holland, "DOE Announces $4 Billion in 48C Allocations," Crux, April 1, 2024, https://www.cruxclimate.com/insights/doe-announces-4-billion-in-48c-allocations.

195 "Applicant Self-Disclosed 48C Projects," U.S. Department of Energy, accessed December 10, 2024, https://www.energy.gov/mesc/applicant-self-disclosed-48c-projects.

196 Joelle Anselmo, "$6B in Inflation Reduction Act Tax Credits for Advanced Energy Projects Open This Month," Utility Dive, May 3, 2024, https://www.utilitydive.com/news/treasury-department-issues-second-round-6-billion-48c-clean-energy-tax-credits/715113/.

197 "Qualifying Advanced Energy Project Credit (§48C) Frequently Asked Questions," U.S. Department of Energy, updated June 10, 2024, https://www.energy.gov/sites/default/files/2024-06/48C_FAQs_06-10-2024.pdf.

198 Assistant Secretary of Defense, Industrial Base Policy, "Defense Production Act III," Office of the Assistant Secretary of Defense Industrial Base Policy, accessed December 1, 2024, https://www.businessdefense.gov/ibr/mceip/dpai/dpat3/docs/DPA-TitleIII-Overview.pdf.

199 Douglass I. Bell, "'A Little-Known Bill of Great National Significance': The Uses and Evolution of the Defense Production Act, 1950-2020," U.S. Army War College, accessed December 1, 2024, https://ahec.armywarcollege.edu/documents/Defense_Production_Act_1950-2020.pdf.

200 U.S. Department of Defense, "DOD Announces Rare Earth Element Awards to Strengthen Domestic Industrial Base,"

press release, November 17, 2020, https://www.defense.gov/News/Releases/Release/article/2418542/dod-announces-rare-earth-element-awards-to-strengthen-domestic-industrial-base/; and Assistant Secretary of Defense, "DoD Awards Key Contract for Domestic Heavy Rare Earth Separation Capability," Office of Industrial Base Policy, June 14, 2022, https://www.businessdefense.gov/news/2022/DoD-Awards-Key-Contract-for-Domestic-Heavy-Rare-Earth-Separation-Capability.html.

201 "Advanced Transportation Financing," U.S. Department of Energy, accessed December 10, 2024, https://www.energy.gov/lpo/advanced-transportation-financing.

202 "LPO's Loans and Loan Guarantees: Overview and Characteristics of Its Financing Options," U.S. Department of Energy, accessed December 10, 2024, https://www.energy.gov/lpo/articles/lpos-loans-and-loan-guarantees-overview-and-characteristics-its-financing-options.

203 "DOE Announces First Advanced Technology Vehicles Manufacturing Loan in More than a Decade," U.S. Department of Energy, July 27, 2022, https://www.energy.gov/articles/doe-announces-first-advanced-technology-vehicles-manufacturing-loan-more-decade.

204 "LPO Offers First Conditional Commitment for Critical Materials Project for Syrah Vidalia to Support Domestic EV Supply Chain," U.S. Department of Energy, April 18, 2022, https://www.energy.gov/lpo/articles/lpo-offers-first-conditional-commitment-critical-materials-project-syrah-vidalia.

205 MINING.COM. "Syrah Resources Declares Force Majeure for Graphite Mine in Mozambique." Accessed December 26, 2024. https://www.mining.com/web/syrah-resources-declares-force-majeure-for-its-graphite-mine-in-mozambique/.

206 "E&E News: Mining Firm Defaults on US Loans amid Mozambique Political Unrest." Accessed December 26, 2024. https://subscriber.politicopro.com/article/eenews/2024/12/12/mining-firm-defaults-on-us-loans-amid-mozambique-political-unrest-00194085.

207 Discussions with the U.S. Department of Energy's Loan Programs Office in February and April 2024.

208 "Copper Ore," Observatory of Economic Complexity, accessed December 10, 2024, https://oec.world/en/profile/hs/copper-ore.

209 "U.S. Energy-Related Carbon Dioxide Emissions, 2023," U.S. Energy Information Administration, April 25, 2024, https://www.eia.gov/environment/emissions/carbon/.

210 Denise Chow, "Led by the South, Clean Energy Sector Added More than 150,000 Jobs Last Year," NBC News, September 18, 2024, https://www.nbcnews.com/science/science-news/clean-energy-job-growth-inflation-reduction-act-rcna171665.

211 Amrith Ramkumar and Phred Dvorak, "The Biggest Winners in America's Climate Law: Foreign Companies," *Wall Street Journal*, July 20, 2023, https://www.wsj.com/articles/the-biggest-winners-in-americas-climate-law-foreign-companies-c672e09d.

212 "Critical Minerals Agreement with Japan Blasted by Congress," Sandler, Travis & Rosenberg, P.A, March 31, 2023, https://www.strtrade.com/trade-news-resources/str-trade-report/trade-report/march/critical-minerals-agreement-with-japan-blasted-by-congress.

213 "Free Trade Agreements," Office of the United States Trade Representative, accessed December 10, 2024, https://ustr.gov/trade-agreements/free-trade-agreements.

214 "Mineral Commodity Summaries: Rare Earths," United States Geological Survey, January 2024, https://pubs.usgs.gov/periodicals/mcs2024/mcs2024-rare-earths.pdf

215 Cecilia Jamasmie, "Serra Verde Makes Global List of Critical Projects," Mining.com, October 22, 2024, https://www.mining.com/serra-verde-makes-global-list-of-critical-projects.

216 S&P Global Market Intelligence, "Metals & Mining," Capital IQ Pro (database), accessed December 10, 2024, https://www.capitaliq.spglobal.com/web/client?.

217 "Platinum-Group Metals, Mineral Commodity Summaries 2024," U.S. Geological Survey. January 2024, https://pubs.usgs.gov/periodicals/mcs2024/mcs2024-platinum-group.pdf.

218 S&P Global Market Intelligence, "Metals & Mining."

219 Reuters, "Sibanye's Montana Woes Underscore Miners' Growing Reliance on Washington," Mining.com, October 23, 2024, https://www.mining.com/web/sibanyes-montana-woes-underscore-miners-growing-reliance-on-washington.

220 Francisco A. Moris, "Semiconductors: the building blocks of the information revolution," *Monthly Labor Review*, Bureau of Labor Statistics, August 1996, https://www.bls.gov/opub/mlr/1996/08/art2full.pdf.

221 Rose Marie Ham, Greg Linden, and Melissa M. Appleyard, "The Evolving Role of Semiconductor Consortia in the United States and Japan," *California Management Review* 41, no. 1 (Fall 1998): 137-163, https://pdxscholar.library.pdx.edu/cgi/viewcontent.cgi?article=1022&context=busadmin_fac.

222 Emily G. Belvins, Alice B. Grossman, and Karen M. Sutter, *Frequently Asked Questions: CHIPS Act of 2022 Provisions and Implementation*, CRS Report No. R47523 (Washington, DC: Congressional Research Service, April 2023), https://crsreports.congress.gov/product/pdf/R/R47523.

223 "Taiwan's dominance of the chip industry makes it more important," *The Economist*, March 6, 2023, https://www.economist.com/special-report/2023/03/06/taiwans-dominance-of-the-chip-industry-makes-it-more-important.

224 Don Clark, "'It's a Roller-Coaster Ride': Global Chip Shortage Is Making Industries Sweat," *New York Times*, April 15, 2021, https://www.nytimes.com/2021/04/15/technology/computer-chip-semiconductor-shortage.html.

225 Laura Weiss, "'Chips-plus' bill advances toward final passage in Senate," Roll Call, July 26, 2022, https://rollcall.com/2022/07/26/chips-plus-bill-advances-toward-final-passage-

in-senate/.

226 "Executive Order on America's Supply Chains," The White House, February 24, 2021, https://www.whitehouse.gov/briefing-room/presidential-actions/2021/02/24/executive-order-on-americas-supply-chains/.

227 "The Microelectronics Commons: A National Network of Prototyping Innovation Hubs," Microelectronics Commons, accessed October 9, 2024, https://microelectronicscommons.org/.

228 *CHIPS and Science Act of 2022*, H.R.4346 - 117th Congress (2021-2022), https://www.congress.gov/bill/117th-congress/house-bill/4346/.

229 U.S. Department of State, "Funding for International Partnerships Through the CHIPS Act," press briefing, March 15, 2023, https://www.state.gov/briefings-foreign-press-centers/funding-for-international-partnerships-through-the-chips-act.

230 *CHIPS and Science Act of 2022.*

231 "Minerals Security Partnership," U.S. Department of State, accessed October 9, 2024, https://www.state.gov/minerals-security-partnership/.

232 "Minerals Security Partnership (MSP) Principles for Responsible Critical Mineral Supply Chains," U.S. Department of State, February 2023, https://www.state.gov/wp-content/uploads/2023/02/MSP-Principles-for-Responsible-Critical-Mineral-Supply-Chains-Accessible.pdf.

233 CHIPS for America, Vision for Success: *Facilities for Semiconductor Materials and Manufacturing Equipment* (Washington, DC: National Institute of Standards and Technology, U.S. Department of Commerce, 2023), 6, https://www.nist.gov/system/files/documents/2023/09/29/Vision%20for%20Success-Facilities%20for%20Semiconductor%20Materials%20and%20Mfr%20Equipment.pdf.

234 CHIPS for America, "Fact Sheet: Dedicated Supply Chain Notice of Funding Opportunity," National Institute of Standards and Technology, U.S. Department of Commerce, September 29, 2023, https://www.nist.gov/system/files/documents/2023/09/29/Notice%20of%20Funding%20Opportunity%20Small-Scale%20Supplier%20Projects%20Fact%20Sheet.pdf.

235 CHIPS for America, *Vision for Success: Commercial Fabrication Facilities* (Washington, DC: National Institute of Standards and Technology, U.S. Department of Commerce, 2023), 6, https://www.nist.gov/system/files/documents/2023/02/28/Vision_for_Success-Commercial_Fabrication_Facilities.pdf.

236 Ibid., 8.

237 GlobalFoundries, "GlobalFoundries and Biden-Harris Administration Announce CHIPS and Science Act Funding for Essential Chip Manufacturing," press release, February 19, 2024, https://gf.com/gf-press-release/globalfoundries-and-biden-harris-administration-announce-chips-and-science-act-funding-for-essential-chip-manufacturing/.

238 "Awards Made in First Round of Microelectronics Commons Call for Projects, Totaling Nearly $269M," National Security Technology Accelerator, September 17, 2024, https://nstxl.org/awards-made-in-first-round-of-microelectronics-commons-projects/.

239 Le Lézard, "The MMEC Receives Microelectronics Commons Project Awards," press release, September 19, 2024, https://www.lelezard.com/en/news-21532116.html.

240 Northeast Microelectronics Coalition Hub, "Healey-Driscoll Administration Celebrates Nearly $38 Million Federal Award to Spur Microelectronics Development," Massachusetts Technology Collaborative, September 18, 2024, https://masstech.org/news/healey-driscoll-administration-celebrates-nearly-38-million-federal-award-spur.

241 Matthew P. Funaiole, Brian Hart, and Aidan Powers-Riggs, *De-risking Gallium Supply Chains: The National Security Case for Eroding China's Critical Mineral Dominance* (Washington, DC: CSIS, August 2023), https://www.csis.org/analysis/de-risking-gallium-supply-chains-national-security-case-eroding-chinas-critical-mineral.

242 "The semiconductor industry faces its biggest technical challenge yet," *The Economist*, September 16, 2024, https://www.economist.com/technology-quarterly/2024/09/16/the-semiconductor-industry-faces-its-biggest-technical-challenge-yet.

243 Aaron D. Franklin, Mark C. Hersam, and H.S. Philip Wong, "Carbon nanotube transistors: Making electronics from molecules," *Science* 378, no. 6621 (November 2022), 726-732, https://www.science.org/doi/10.1126/science.abp8278.

244 Yanfang Huang et al., "The Extraction and Separation of Scarce Critical Metals: A Review of Gallium, Indium and Germanium Extraction and Separation from Solid Wastes," *Separations* 11, no. 4 (March 24, 2024), https://doi.org/10.3390/separations11040091.

245 Nora K. Foley, Brian W. Jaskula, Bryn E. Kimball, and Ruth F. Schulte, "Chapter H: Gallium," in *Critical Mineral Resources of the United States–Economic and Environmental Geology and Prospects for Future Supply*, ed. Klaus J. Schulz, John H. DeYoung, Jr., Robert R. Seal II, and Dwight Bradley (Reston, Virginia: U.S. Geological Survey, 2017), https://pubs.usgs.gov/pp/1802/h/pp1802h.pdf.

246 "Executive Order on America's Supply Chains," The White House, February 24, 2021, https://www.whitehouse.gov/briefing-room/presidential-actions/2021/02/24/executive-order-on-americas-supply-chains/.

247 The White House, *Building Resilient Supply Chains, Revitalizing American Manufacturing, and Fostering Broad-Based Growth*, 100-Day Reviews under Executive Order 14017 (Washington, DC: The White House, June 2021), https://www.whitehouse.gov/wp-content/uploads/2021/06/100-day-supply-chain-review-report.pdf.

248 U.S. Department of Defense (DOD), *Securing Defense-Critical Supply Chains* (Washington, DC: DOD, February 2022, https://www.businessdefense.gov/docs/eo/DoD-EO14017-Report_Securing_Defense-Critical-Supply-Chains.pdf.

249 U.S. Department of Defense, *Assessing and Strengthening the Manufacturing and Defense Industrial Base and Supply Chain Resiliency of the United States* (Washington, DC: DOD, September 2018), https://media.defense.gov/2018/Oct/05/2002048904/-1/-1/1/ASSESSING-AND-STRENGTHENING-THE-MANUFACTURING-AND%20DEFENSE-INDUSTRIAL-BASE-AND-SUPPLY-CHAIN-RESILIENCY.PDF; and "Resources," Industrial Base Policy, Assistant Secretary of Defense, accessed October 19, 2024, https://www.businessdefense.gov/resources.html.

250 Aidan Lawson and June Rhee, "Usage of the Defense Production Act throughout history and to combat COVID-19," Program on Financial Stability, Yale School of Management, June 3, 2020, https://som.yale.edu/blog/usage-of-the-defense-production-act-throughout-history-and-to-combat-covid-19.

251 The White House, *Building Resilient Supply Chains*.

252 "Defense Production Act Title III," Office of the Assistant Secretary of Defense Industrial Base Policy, accessed October 19, 2024, https://www.businessdefense.gov/ibr/mceip/dpai/dpat3/docs/DPA-TitleIII-Overview.pdf; and Luke A. Nicastro, "FY2024 NDAA: Defense Industrial Base Policy," Congressional Research Service, IF12221, updated January 8, 2024, https://crsreports.congress.gov/product/pdf/IN/IN12221.

253 The White House, *Building Resilient Supply Chains*.

254 U.S. Department of Defense, "Defense Production Act Title III Presidential Determination for Critical Materials in Large-Capacity Batteries," press release, April 5, 2022, https://www.defense.gov/News/Releases/Release/Article/2989973/defense-production-act-title-iii-presidential-determination-for-critical-materi/; and "History of DPA Title III President Determinations and Waivers," Assistant Secretary of Defense, Industrial Base Policy, accessed November 9, 2024, https://www.businessdefense.gov/ibr/mceip/dpai/dpat3/docs/DPA-TitleIII-Recent-Presidential-Determinations.pdf.

255 Ian Satchwell, "China's control and coercion in critical minerals," *The Strategist*, Australian Strategic Policy Institute, June 7, 2024, https://www.aspistrategist.org.au/chinas-control-and-coercion-in-critical-minerals.

256 "Critical Minerals & Materials Program" U.S. Department of Energy, accessed November 9, 2024, https://www.energy.gov/cmm/critical-minerals-materials-program.

257 "Defense Production Act (DPA) Title III Receives Emergency Supplemental Funding for Ukraine," Industrial Base Policy, Assistant Secretary of Defense, accessed October 19, 2024, https://www.businessdefense.gov/ibr/mceip/dpai/dpat3/docs/announcements/UkraineSupplmentalFundsfortheDPATitleIII-Fund.pdf.

258 Ibid.

259 Noah Robertson, "How a 1950s-era law has become a flashpoint for the US arms industry," Defense News, August 21, 2024, https://www.defensenews.com/pentagon/2024/08/21/how-a-1950s-era-law-has-become-a-flashpoint-for-the-us-arms-industry/.

260 Christopher Zember, senior adviser and portfolio manager, Manufacturing Capability Expansion Pathfinders, Office of the Secretary of Defense, Industrial Base Policy.

261 "DoD's Defense Business Accelerator Speeds Selection, Funding, and Commercialization of Dual-use Technologies," Acquisition & Sustainment, Office of the Under Secretary of Defense, June 7, 2024, https://www.acq.osd.mil/news/archive/2024/DODs-Defense-Business-Accelerator-Speeds-Selection.html.

262 "Manufacturing Capability Expansion and Investment Prioritization," presentation at the National Defense Industrial Association Manufacturing Division meeting, May 2, 2024.

263 Brad Dress, "China will be ready for potential Taiwan invasion by 2027, US admiral warns," *The Hill*, March 21, 2024, https://thehill.com/policy/defense/4547637-china-potential-taiwan-invasion-2027-us-admiral-warns/.

264 "Magnetic Applications for the Automotive Industry," Bunting, accessed October 19, 2024, https://bunting-dubois.com/magnetic-applications-for-the-automotive-industry.

265 U.S. Department of the Interior and U.S. Geological Survey, "A World of Minerals in Your Mobile Device," USGS, General Information Product 167, September 2016, https://pubs.usgs.gov/gip/0167/gip167.pdf.

266 The White House, *Building Resilient Supply Chains, Revitalizing American Manufacturing, and Fostering Broad-Based Growth* (Washington, DC: The White House, 2021), 11, https://www.whitehouse.gov/wp-content/uploads/2021/06/100-day-supply-chain-review-report.pdf; and U.S. Department of Commerce, *A Federal Strategy to Ensure Secure and Reliable Supplies of Critical Minerals* (Washington, DC: U.S. Department of Commerce, 2019), 27, https://www.commerce.gov/sites/default/files/2020-01/Critical_Minerals_Strategy_Final.pdf.

267 As of early October 2024, the MSP partners are Australia, Canada, Estonia, Finland, France, Germany, India, Italy, Japan, Norway, South Korea, Sweden, the United Kingdom, the United States, and the European Union. Since July 1, 2024, South Korea is the MSP chair. See "Minerals Security Partnership," U.S. Department of State, accessed October 11, 2024, https://www.state.gov/minerals-security-partnership/.

268 Ibid.

269 Ibid.

270 Hannah Northey, "The Diplomat Securing Responsibly Sourced Critical Minerals," *Politico*, January 25, 2023, https://www.politico.com/newsletters/the-long-game/2023/01/25/the-diplomat-securing-responsibly-sourced-critical-minerals-00079397; and U.S. Department of State, "Minerals Security Partnership Convening Supports Robust Supply Chains for Clean Energy Technologies," press release, September 22, 2022, https://www.state.gov/minerals-security-partnership-convening-supports-robust-supply-chains-for-clean-energy-technologies.

271 U.S. Department of State, "Minerals Security Partnership

Principles for Responsible Critical Mineral Supply Chains," U.S. Department of State, February 2023, https://www.state.gov/wp-content/uploads/2023/02/MSP-Principles-for-Responsible-Critical-Mineral-Supply-Chains-Accessible.pdf.

272 Charles Chang et al., "China's Global Research Grows behind Critical Minerals," S&P Global, August 24, 2023, https://www.spglobal.com/content/dam/spglobal/corporate/en/images/general/special-editorial/083123-china-s-global-reach-grows-behind-critical-minerals.pdf.

273 Northey, "The Diplomat Securing."

274 Hannah Northey, "U.S. and Its Partners Unveil Plan for Financing Minerals Projects," POLITICOPro, September 24, 2024, https://subscriber.politicopro.com/article/2024/09/us-and-its-partners-unveil-plans-for-financing-minerals-projects-pro-00180483.

275 U.S. Department of State, "Joint Statement of the Minerals Security Partnership," press release, March 4, 2024, https://www.state.gov/joint-statement-of-the-minerals-security-partnership/.

276 U.S. Department of State, "Joint Statement on Establishment of the Minerals Security Partnership Finance Network," press release, September 23, 2024, https://www.state.gov/joint-statement-on-establishment-of-the-minerals-security-partnership-finance-network/.

277 U.S. Department of State, "Joint Statement of the Minerals Security Partnership."https://www.state.gov/joint-statement-of-the-minerals-security-partnership/

278 U.S. Department of State, "Statement on the First High-Level Minerals Security Partnership (MSP) Forum Event," press release, July 18, 2024, https://www.state.gov/statement-on-the-first-high-level-minerals-security-partnership-msp-forum-event/.

279 U.S. Department of State, "Secretary Blinken Officially Launches the Minerals Security Partnership Forum with the European Commission," press release, April 5, 2024, https://www.state.gov/secretary-blinken-officially-launches-the-minerals-security-partnership-forum-with-the-european-commission/.

280 U.S. Department of State, "Joint Statement on Establishment."

281 Lauren Herzer Risi and Claire Doyle, "Examining China's Impact on Mining in Africa: Critiques and Credible Responses," Wilson Center, July 18, 2023, https://www.wilsoncenter.org/blog-post/examining-chinas-impact-mining-africa-critiques-and-credible-responses.

282 Business and Human Rights Resource Center, *Unpacking Clean Energy: Human Rights Impacts of Chinese Overseas Investment in Transition Minerals* (London: Business and Human Rights Resource Center, 2023), https://media.business-humanrights.org/media/documents/2023_China_TM_briefing.pdf.

283 U.S. Department of State, "Minerals Security Partnership Principles."

284 SAFE Center for Critical Minerals Strategy, *A Global Race to the Top: Using Transparency to Secure Critical Mineral Supply Chains* (Washington, DC: SAFE, 2023), 45, https://safe2020.wpenginepowered.com/wp-content/uploads/2023/03/SAF-_CritMinReport_v06.3_Spreads_Final.pdf.

285 "Minerals Security Partnership," U.S. Department of State.

286 U.S. Congress, Senate, *Better Utilization of Investments Leading to Development Act of 2018 or the BUILD Act of 2018*, S. 2463, 115th Cong., 2nd sess., introduced in Senate February 27, 2018, https://www.congress.gov/bill/115th-congress/senate-bill/2463.

287 "Our Products," U.S. International Development Finance Corporation, accessed October 11, 2024, https://www.dfc.gov/what-we-offer/our-products.

288 Shayerah I. Akhtar and Nick M. Brown, "U.S. International Development Finance Corporation (DFC)," Congressional Research Service, IF11436, updated June 1, 2023, https://crsreports.congress.gov/product/pdf/IF/IF11436.

289 U.S. Geological Survey (USGS), *Mineral Commodity Summaries 2024* (Washington, DC: USGS, 2024), https://pubs.usgs.gov/periodicals/mcs2024/mcs2024.pdf.

290 "Active DFC Projects - List View," U.S. International Development Finance Corporation, accessed October 14, 2024, https://www3.dfc.gov/DFCProjects.

291 Yvonne Yue Li, "US Gives $50 Million Boost to Critical Minerals Investor TechMet," Bloomberg, December 1, 2023, https://www.bloomberg.com/news/articles/2023-12-01/us-gives-50-million-boost-to-critical-minerals-investor-techmet.

292 U.S. International Development Finance Corporation, "DFC Makes More Than $9.1 Billion in Financial Commitments for Fiscal Year 2023," press release, October 5, 2023, https://www.dfc.gov/media/press-releases/dfc-makes-more-91-billion-financial-commitments-fiscal-year-2023.

293 "Interview: Lack of Mapping, Technologies, Infrastructure Hampers Africa from Exploring Mineral Resources," Xinhua, June 9, 2024, https://english.news.cn/africa/20240609/76175814668d462684e11ba89ead6e3b/c.html.

294 Akhtar and Brown, *U.S. International Development Finance Corporation*. https://crsreports.congress.gov/product/pdf/IF/IF11436

295 U.S. International Development Finance Corporation (DFC), *A Guide to Partnering with U.S. International Development Finance Corporation* (Washington, DC: DFC, 2021), https://www.dfc.gov/sites/default/files/media/documents/DFC_Partner_Guide.pdf.

296 The World Bank, "Overview of State Ownership in the Global Minerals Industry," *Extractive Industries for Development Series 20* (Washington, DC: The World Bank, 2011), 32, https://documents1.worldbank.org/curated/ar/339551468340825224/pdf/828480NWP0Extr00Box379875B00PUBLIC0.pdf.

297 "Overview," Japan Oil, Gas, and Metals National Corporation, accessed October 14, 2024, https://www.jogmec.go.jp/english/about/about001.html.

298 Japan Oil, Gas, and Metals National Corporation, *Business Tool List: An Introduction of Support Tools of the Metals Departments*

(Tokyo: JOGMEC, 2021), 6-8, https://www.jogmec.go.jp/content/300196027.pdf.

299 Japan Oil, Gas, and Metals National Corporation, "Sojitz and JOGMEC Enter into Definitive Agreements with Lynas Including Availability Agreement to Secure Supply of Rare Earths Products to Japanese Market," press release, March 30, 2011, https://www.jogmec.go.jp/english/news/release/release0069.html; and Joseph Rachman, "To Answer China on Rare Earths, the US Should Look to Japan," *Foreign Policy*, August 28, 2023, https://foreignpolicy.com/2023/08/15/japan-rare-earth-minerals-green-transition-china-supply-chains/.

300 Nayan Seth, "How to Diversify Mineral Supply Chains - A Japanese Agency Has Lessons for All," New Security Beat, August 15, 2024, https://www.newsecuritybeat.org/2024/08/how-to-diversify-mineral-supply-chains-a-japanese-agency-has-lessons-for-all/.

301 H. W. Halleck, "Introductory Remarks," in J. H. N. de Fooz, *Fundamental Principles of the Law of Mines*, trans. H. W. Halleck, (San Francisco: J. B. Painter, 1860), v.

302 Ibid., v.

303 Ibid., vi.

304 Mary C. Rabbitt, Minerals, Lands, and Geology for the Common Defence and General Welfare, Volume 1, Before 1879: A History of Geology in Relation to the Development of Public Lands, Federal Science, and Mapping Policy, and the Development of Mineral Resources in the United States (Washington, DC: U.S. Government Printing Office, 1979), 2, https://pubs.usgs.gov/book/1979/rabbitt-vol1/report.pdf; and Clarence King, The United States Mining Laws and Regulations Thereunder, and State and Territorial Mining Laws to Which Are Appended Local Mining Rules and Regulations (Washington, DC: U.S. Government Printing Office, 1885), 1, https://babel.hathitrust.org/cgi/pt?id=uc1.31175021972180&seq=19.

305 Act Granting the Right of Way to Ditch and Canal Owners over the Public Lands, and for Other Purposes, Chap. CCLXII, U.S. Statutes at Large 251 (July 26, 1866): 251-253, https://digitalcommons.csumb.edu/cgi/viewcontent.cgi?article=1013&context=hornbeck_usa_2_d.

306 An Act to Amend "An Act Granting the Right of Way to Ditch and Canal Owners over the Public Lands, and for Other Purposes," Chap. CCXXXV, U.S. Statutes at Large 16 (July 9, 1870): 217-218, https://www.govinfo.gov/content/pkg/STATUTE-16/pdf/STATUTE-16-Pg217.pdf; King, The United States Mining Laws, 9; Marc Humphries, Mining on Federal Lands: Hardrock Minerals, CRS Report No. RL33908 (Washington, DC: Congressional Research Service, updated July 2009), 1, https://crsreports.congress.gov/product/pdf/RL/RL33908.

307 King, The United States Mining Laws, 1; and Humphries, Mining on Federal Lands.

308 Mary C. Rabbitt, Minerals, Lands, and Geology for the Common Defence and General Welfare, Volume 2, 1879-1904: A History of Geology in Relation to the Development of Public-Land, Federal-Science, and Mapping Policies and the Development of Mineral Resources in the United States during the First 25 Years of the U.S. Geological Survey (Washington, DC: U.S. Government Printing Office, 1980), 5, https://pubs.usgs.gov/book/1980/rabbitt-vol2/report.pdf.

309 Nearly 40 years after passage of the Mining Act of 1872, U.S. secretary of the interior Richard A. Ballinger noted, "Granting defects in the laws, they have accomplished their purpose in causing the mineral resources to be developed and have thus contributed enormously to the wealth of the nation." The U.S. Geological Survey later pointed to the 1880s as the start of the U.S. mining boom. See Richard A. Ballinger, Report of the Secretary of the Interior: For the Fiscal Year Ended June 30, 1910 (Washington, DC: U.S. Government Printing Office, 1910), 13, https://babel.hathitrust.org/cgi/pt?id=mdp.39015031655015&seq=19; and George Otis Smith, "The Public Interest in Mineral Resources," in Mineral Resources of the United States 1915, Part 1, ed. H. D. McCaskey (Washington, DC: U.S. Government Printing Office, 1917), 3a, https://babel.hathitrust.org/cgi/pt?id=uc1.$b570202&seq=13.

310 George Otis Smith, "The Public Interest in Mineral Resources," in *Mineral Resources of the United States 1915* 1, ed. H. D. McCaskey (Washington, DC: U.S. Government Printing Office, 1917), 9a, https://babel.hathitrust.org/cgi/pt?id=uc1.$b570202&seq=19. In 1915, George Otis Smith, director of the U.S. Geological Survey, wrote, "The governmental duty to the mining industry first of all is to promote use without waste." See George Otis Smith, "The Public Interest in Mineral Resources," in *Mineral Resources of the United States 1915* 1, ed. H. D. McCaskey (Washington, DC: U.S. Government Printing Office, 1917), 7a, https://babel.hathitrust.org/cgi/pt?id=uc1.$b570202&seq=17.

311 Smith, "The Public Interest in Mineral Resources," 8a.

312 Mary C. Rabbitt, *Minerals, Lands, and Geology for the Common Defence and General Welfare, Volume 3, 1904-1939: A History of Geology in Relation to the Development of Public-Land, Federal-Science, and Mapping Policies and the Development of Mineral Resources in the United States from the 25th to the 60th Year of the U.S. Geological Survey* (Washington, DC: U.S. Government Printing Office, 1986), 195, https://pubs.usgs.gov/book/1986/rabbitt-vol3/report.pdf. In 1917, Congress did pass a bill that opened some public lands to potash prospecting and development.

313 Ibid., 300, 347-348.

314 Ibid., 368.

315 H. Herbert Hughes, "Introduction," in *Minerals Yearbook 1939*, ed. H. Herbert Hughes (Washington, DC: U.S. Government Printing Office, 1939), ix, https://search.library.wisc.edu/digital/AQGHPBQDKJTJZH9E/pages/AEBUQZMODVA3VD8F; and E. W. Pehrson, "Review of the Mineral Industries," in *Minerals Yearbook 1941*, ed. F. M. Shore (Washington, DC: U.S. Government Printing Office, 1943), xxii-xxiii, https://search.library.wisc.edu/digital/AVZLKWT7LNXCK39B/pages/ARXACYKDZMWMFY8J. For more information on federal aids for mineral production during World War II, see U.S. War Production

Board, *Federal Aids for War Mineral Production*, 2d ed. (Washington, DC: War Production Board, December 1942), https://babel.hathitrust.org/cgi/pt?id=uiug.30112069635347&seq=1.

316 E. W. Pehrson and H. D. Keiser, "Review of the Mineral Industries in 1940," in *Minerals Yearbook Review of 1940*, ed. H. D. Keiser (Washington, DC: U.S. Government Printing Office, 1941), 1, https://search.library.wisc.edu/digital/ANL3GAHUQMHTGL8Q/pages/A3FXV4YMB4KIZ687.

317 "What Is the Authority for This Part," *Code of Federal Regulations*, Title 43 (2005): 3501.1, https://www.ecfr.gov/current/title-43/subtitle-B/chapter-II/subchapter-C/part-3500; Anne-Marie Fennell, "Mining on Federal Lands: More than 800 Operations Authorized to Mine and Total Mineral Production Is Unknown," GAO-20-461R, U.S. Government Accountability Office, May 28, 2020, footnote 6, https://www.gao.gov/assets/gao-20-461r.pdf; and Humphries, *Mining on Federal Lands*, 3.

318 John D. Morgan, Jr., "National Stockpile and United States Strategy," speech at Industrial College of the Armed Forces, Washington, DC, December 6, 1955, https://www.hsdl.org/?view&did=456593; and Harry S. Truman, "Annual Budget Message to the Congress: Fiscal Year 1952," Harry S. Truman Library & Museum, January 15, 1951, https://www.trumanlibrary.gov/library/public-papers/13/annual-budget-message-congress-fiscal-year-1952.

319 U.S. Government Accountability Office (GAO), *Federal Land Management: Key Differences and Stakeholder Views of the Federal Systems Used to Manage Hardrock Mining*, GAO-21-299 (Washington, DC: GAO, July 2021, 6, https://www.gao.gov/assets/gao-21-299.pdf.

320 The Resources and Conservation Act was introduced in 1959, but it failed to pass. See *Proposed Resources and Conservation Act of 1960: Hearing on S. 2549 before the Committee on Interior and Insular Affairs, United States Senate*, 86th Cong., 2d sess. (January 25, 1960) (statement of James E. Murray, U.S. Senator from Montana), 4, https://babel.hathitrust.org/cgi/pt?id=umn.31951d02087210h&seq=12; and Dinah Bear, "The National Environmental Policy Act: Its Origins and Evolutions," *Natural Resources & Environment* 10, no. 2 (1995): 3, https://www.jstor.org/stable/40923443.

321 For example, the Expanded Wilderness Act in 1964 made special provisions for mineral extraction in national wilderness areas. *See An Act to Establish a National Wilderness Preservation System for the Permanent Good of the Whole People, and for Other Purposes*, Public Law 88-577, *U.S. Statutes at Large 78* (September 3, 1964): 894-896, https://www.govinfo.gov/content/pkg/STATUTE-78/pdf/STATUTE-78-Pg890.pdf; and Olman Hee and Jeannette I. Baker, "Review of the Mineral Industries," in *Minerals Yearbook 1968: Metals, Minerals, and Fuels* 1-2, ed. Albert E. Schreck(Washington, DC: U.S. Government Printing Office, 1969), 15, https://search.library.wisc.edu/digital/AMNS3JOQHHHFEG9C/pages/A462ANUJ4BVF5U8D.

322 Bernadette S. Schumaker and Jeannette I. Baker, "Review of the Mineral Industries," in Minerals Yearbook 1969: *Metals, Minerals, and Fuels* 1-2, ed. Albert E. Schreck (Washington, DC: U.S. Government Printing Office, 1969), 17, https://search.library.wisc.edu/digital/AXGOZDSUNZXVBA8V/pages/ANIMZQYAOYT3G28X; and Bernadette S. Schumaker and Jeannette I. Baker, "Review of the Mineral Industries," in *Minerals Yearbook 1970: Metals, Minerals, and Fuels*, Part 1, ed. Albert E. Schreck (Washington, DC: U.S. Government Printing Office, 1972), 17, https://search.library.wisc.edu/digital/ADV7JLBE7QKH6C9C/pages/AO4HNL7TRCQ5TQ9A.

323 Bernadette S. Schumaker and Jeannette I. Baker, "Review of the Mineral Industries," in *Minerals Yearbook 1969: Metals, Minerals, and Fuels*, Parts 1-2, ed. Albert E. Schreck (Washington, DC: U.S. Government Printing Office, 1969), 17, https://search.library.wisc.edu/digital/AXGOZDSUNZXVBA8V/pages/ANIMZQYAOYT3G28X; and Schumaker and Baker, "Review of the Mineral Industries," 17.

324 *An Act to Establish a National Policy for the Environment, to Provide for the Establishment of a Council on Environmental Quality, and for Other Purposes, Public Law* 91-190, U.S. Statutes at Large 83 (January 1, 1970): 853, https://www.govinfo.gov/content/pkg/STATUTE-83/pdf/STATUTE-83-Pg852.pdf#page=1.

325 John L. Morning, "Technologic Trends in the Mineral Industries (Metals and Nonmetals except Fuels)," in *Minerals Yearbook 1970*, 62; and John L. Morning, "Technologic Trends in the Mineral Industries (Metals and Nonmetals except Fuels)," in *Minerals Yearbook 1971: Metals, Minerals, and Fuels*, Part 1, ed. Albert E. Schreck (Washington, DC: U.S. Government Printing Office, 1973), 60, https://search.library.wisc.edu/digital/ARFPBSJPN63JHT86/pages/APODKCXHBPYNJA9D.

326 Morning, "Technologic Trends in the Mineral Industries," 57.

327 John L. Morning, "Technologic Trends in the Mineral Industries (Metals and Nonmetals except Fuels)," in *Minerals Yearbook 1972: Metals, Minerals, and Fuels*, Part 1, ed. Albert E. Schreck (Washington, DC: U.S. Government Printing Office, 1974), 61, 63, https://search.library.wisc.edu/digital/A3NTU4IUYRMAY48H/pages/AB4G76U2SHNBBP8W.

328 Daniel L. Sullivan et al., "Review of the Mineral Industries," in *Minerals Yearbook 1972*, 20-21; and *An Act to Amend the Federal Water Pollution Control Act*, Public Law 92-500, *U.S. Statutes at Large* 86 (October 18, 1972): 816-903, https://www.govinfo.gov/content/pkg/STATUTE-86/pdf/STATUTE-86-Pg816.pdf.

329 John L. Morning, "Mining and Quarrying Trends in the Metal and Nonmetal Industries," in *Minerals Yearbook 1973: Metals, Minerals, and Fuels*, Part 1, ed. Albert E. Schreck (Washington, DC: U.S. Government Printing Office, 1975), 67, https://search.library.wisc.edu/digital/AKPKKOK4OMFJ3O8Q/pages/ANHMAMMNIZSMZH8U; Daniel L. Sullivan and Nicholas G. Theofilos, "Review of the Mineral Industries," in *Minerals Yearbook 1974: Metals, Minerals, and Fuels*, Part 1, ed. Albert E. Schreck (Washington, DC: U.S. Government Printing Office, 1976), 2, https://search.library.wisc.edu/digital/AET6LTJXWL7TNK8M/pages/A75AUQRNEL5AQ28Y; John L. Morning, "Mining and Quarrying Trends in the Metal and Nonmetal Industries," in *Minerals Yearbook 1975: Metals, Minerals, and*

Fuels, Part 1, ed. Albert E. Schreck (Washington, DC: U.S. Government Printing Office, 1977), 72, https://search.library.wisc.edu/digital/AZEG44QN6JCYZJ9A/pages/AZV5YVG27DL7IZ8S; Barry W. Klein et al., "Review of the Mineral Industries," in *Minerals Yearbook 1976: Metals, Minerals, and Fuels*, Part 1, ed. Albert E. Schreck (Washington, DC: U.S. Government Printing Office, 1978), 2, https://search.library.wisc.edu/digital/A7YV676KA3AHUK8Z/pages/ACZOX7B5YBW7BU86; Barry W. Klein et al., "Review of the Mineral Industries," in *Minerals Yearbook 1977: Metals and Minerals*, Part 1, ed. Albert E. Schreck (Washington, DC: U.S. Government Printing Office, 1980), 11-12, https://search.library.wisc.edu/digital/A4QWEORRX53TWH8K/pages/ASV6OXA2CEXKCW8V; Franklin D. Cooper, "Mining and Quarrying Trends in the Metal and Nonmetal Industries," in *Minerals Yearbook 1978-79: Metals and Minerals*, Part 1, ed. Albert E. Schreck (Washington, DC: U.S. Government Printing Office, 1980), 1, https://search.library.wisc.edu/digital/ALAGORVJYOGFX28A/pages/AOTTLYAB5ZDFEH9E; and Comptroller General of the United States, "The U.S. Mining and Mineral-Processing Industry: An Analysis of Trends and Implications," ID-80-04, U.S. General Accounting Office, October 31, 1979, ii, iv, https://www.gao.gov/assets/id-80-04.pdf. For example, according to the U.S. Bureau of Mines in 1974, "The nonferrous metals industry had the largest expenditure on pollution control equipment (as a proportion of total capital expenditures) of any industry in the United States." See Daniel L. Sullivan and Nicholas G. Theofilos, "Review of the Mineral Industries," in *Minerals Yearbook 1974*, 2.

330 T. A. Johnson, "Mining and Quarrying Trends in the Metal and Nonmetal Industries," in Minerals Yearbook 1976, 71.

331 Daniel E. Sullivan et al., "Review of the Mineral Industries," in Minerals Yearbook 1975, 2; and Franklin D. Cooper, "Mining and Quarrying Trends in the Metal and Nonmetal Industries," in *Minerals Yearbook 1978-79: Metals and Minerals*, Part 1, ed. Albert E. Schreck (Washington, DC: U.S. Government Printing Office, 1980), 1, https://search.library.wisc.edu/digital/ALAGORVJYOGFX28A/pages/AOTTLYAB5ZDFEH9E.

332 Johnson, "Mining and Quarrying Trends in the Metal and Nonmetal Industries," 71.

333 Comptroller General of the United States, *The U.S. Mining and Mineral-Processing Industry: An Analysis of Trends and Implications*, ID-80-04 (Washington, DC: GAO, October 1979), 23, https://www.gao.gov/assets/id-80-04.pdf.

334 Michael Bennon and Devon Wilson, "NEPA Litigation over Large Energy and Transport Infrastructure Projects," *Environmental Law Reporter* 53, no. 10 (October 2, 2023): 10839, https://ssrn.com/abstract=4498938.

335 Brian T. Brady et al., "Mining and Quarrying Trends in the Metals and Industrial Metals Industries," in *Minerals Yearbook 1991*, Part 1, ed. Donald G. Rogich (Washington, DC: U.S. Government Printing Office, 1991), 47, https://search.library.wisc.edu/digital/A5X7AVV22D2URO8R/pages/AJRIKKNOIEN4UV8P.

336 GAO, *Critical Minerals: Status, Challenges, and Policy Options for Recovery from Nontraditional Sources*, GAO-24-106395 (Washington, DC: GAO, July 2024, 25, https://www.gao.gov/assets/gao-24-106395.pdf.

337 Environmental Protection Agency, "Mining Issues: EPA Region 9 Regional Tribal Operations Committee Meeting," presentation, April 20, 2023, slide 13, https://www.epa.gov/system/files/documents/2023-04/r9-rtoc-presentation-mining-issues-breakout-spring-2023.pdf.

338 Examples of federal programs with permitting authority delegated to state governments include section 401 certifications under the Clean Water Act and Underground Injection Control permits under the Safe Drinking Water Act. See Environmental Protection Agency, "Mining Issues: EPA Region 9 Regional Tribal Operations Committee Meeting," presentation, April 20, 2023, slide 17-18, https://www.epa.gov/system/files/documents/2023-04/r9-rtoc-presentation-mining-issues-breakout-spring-2023.pdf.

339 "NPDES Program Authorizations (as of July 2019)," Environmental Protection Agency, February 2021, https://www.epa.gov/sites/default/files/2021-02/documents/authorized_states_2021.pdf; Environmental Protection Agency, "Mining Issues: EPA Region 9 Regional Tribal Operations Committee Meeting," presentation, April 20, 2023, slide 18, https://www.epa.gov/system/files/documents/2023-04/r9-rtoc-presentation-mining-issues-breakout-spring-2023.pdf; and Interagency Working Group on Mining Laws, Regulations, and Permitting, *Recommendations to Improve Mining on Public Lands* (Washington, DC: U.S. Department of the Interior, September 2023), 50, https://www.doi.gov/sites/default/files/mriwg-report-final-508.pdf.

340 Environmental Protection Agency, "Mining Issues: EPA Region 9 Regional Tribal Operations Committee Meeting," presentation, April 20, 2023, slide 12, https://www.epa.gov/system/files/documents/2023-04/r9-rtoc-presentation-mining-issues-breakout-spring-2023.pdf. For an example of a prospective mine requiring an NPDES permit, see Talon Metals Corp. and DRA Americas Inc., "NI 43-101 Technical Report Preliminary Economic Assessment (PEA) #3 of the Tamarack North Project - Tamarack, Minnesota," G4932-RPT-01, January 8, 2021, 300-301, https://talonmetals.com/wp-content/uploads/2021/02/Talon-Tamarack-PEA3_2021.pdf.

341 "About NPDES," Environmental Protection Agency, accessed September 19, 2024, https://www.epa.gov/npdes/about-npdes.

342 Environmental Protection Agency, "Mining Issues: EPA Region 9 Regional Tribal Operations Committee Meeting," presentation, April 20, 2023, slide 14, https://www.epa.gov/system/files/documents/2023-04/r9-rtoc-presentation-mining-issues-breakout-spring-2023.pdf; and Interagency Working Group on Mining Laws, Regulations, and Permitting, *Recommendations to Improve Mining on Public Lands*, 50.

343 Environmental Protection Agency, "Mining Issues: EPA Region 9 Regional Tribal Operations Committee Meeting," presentation, April 20, 2023, slide 14, https://www.epa.gov/system/files/documents/2023-04/r9-rtoc-presentation-mining-issues-breakout-spring-2023.pdf; and Interagency Working

Group on Mining Laws, *Regulations, and Permitting, Recommendations to Improve Mining on Public Lands*, 50.

344 SNL Metals & Mining, "Permitting, Economic Value and Mining in the United States," prepared for the National Mining Association, June 19, 2015, 25, https://nma.org/wp-content/uploads/2016/09/SNL_Permitting_Delay_Report-Online.pdf; and "Water," National Mining Association, accessed September 19, 2024, https://nma.org/category/water/.

345 "Water," National Mining Association, accessed September 19, 2024, https://nma.org/category/water/; and Environmental Protection Agency, "Mining Issues: EPA Region 9 Regional Tribal Operations Committee Meeting," presentation, April 20, 2023, slide 10, https://www.epa.gov/system/files/documents/2023-04/r9-rtoc-presentation-mining-issues-breakout-spring-2023.pdf. Notably, the U.S. Supreme Court circumscribed the definition of adjacent wetlands in Sackett vs. Environmental Protection Agency (2023). See *Sackett v. Environmental Protection Agency*, 598 U.S. 1 (2023), https://www.supremecourt.gov/opinions/22pdf/21-454_4g15.pdf; and "*Sackett* Decision Provides Clarity, Substantially Restricts Clean Water Act Jurisdiction Scope," Holland & Knight, May 26, 2023, https://www.hklaw.com/en/insights/publications/2023/05/sackett-decision-provides-clarity-substantially.

346 "U.S. Interactive Map of State and Tribal Assumption under CWA Section 404," Environmental Protection Agency, accessed September 19, 2024, https://www.epa.gov/cwa404g/us-interactive-map-state-and-tribal-assumption-under-cwa-section-404. Michigan and New Jersey have primacy over issuing Clean Water Act Section 404 permits, and other states like Alaska are seeking to assume primacy. See Hannah Northey and Jael Holzman, "States' Water Permitting Push Sparks Mine Concerns," E&E News, February 14, 2024, https://www.eenews.net/articles/states-water-permitting-push-sparks-mine-concerns/.

347 "Clean Water Act: Section 404(c) 'Veto Authority'," Environmental Protection Agency, March 2016, https://www.epa.gov/sites/default/files/2016-03/documents/404c.pdf; Environmental Protection Agency, "Final Determination to Prohibit the Specification of and Restrict the Use for Specification of Certain Waters Within Defined Areas as Disposal Sites; Pebble Deposit Area, Southwest Alaska," *Federal Register* 88, no. 23 (February 3, 2023): 7441-7443, https://www.govinfo.gov/content/pkg/FR-2023-02-03/pdf/2023-02287.pdf. The U.S. Army Corps of Engineers also denied the Pebble mine project under section 404. See Alaska District, U.S. Army Corps of Engineers, "Review of the Application by Pebble Limited Partnership (POA-2017- 00271) in Light of the Prohibitions and Restrictions Imposed by the *Final Determination of the U.S. Environmental Protection Agency Pursuant to Section 404(c) of the Clean Water Act - Pebble Deposit Area, Southwest Alaska (January 2023)*," Record of Decision, CEPOA-ZA, April 15, 2024, https://www.poa.usace.army.mil/Portals/34/docs/regulatory/specialpns/2024/POA-2017-00271.20240415.RecordofDecision.pdf?ver=QRA0ZnafUF28lJwzGqhnqA%3d%3d.

348 Talon Metals Corp. and DRA Americas Inc., "NI 43-101 Technical Report Preliminary Economic Assessment (PEA) #3 of the Tamarack North Project - Tamarack, Minnesota," G4932-RPT-01, January 8, 2021, 297, 302, https://talonmetals.com/wp-content/uploads/2021/02/Talon-Tamarack-PEA3_2021.pdf; Talon Metals Corp. and DRA Americas Inc., "NI 43-101 Technical Report Updated Preliminary Economic Assessment (PEA) of the Tamarack North Project - Tamarack, Minnesota," G4314-RPT-01, March 12, 2020, 253, https://waterlegacy.org/wp-content/uploads/2020/08/Tamarack-Prelim.-Economic-Assessment-NI-43-101-2020.pdf; Walla Walla District, U.S. Army Corps of Engineers, "Reference Guide for U.S. Army Corps of Engineers (Corps) Clean Water Act Section 404 & Rivers and Harbors Act Section 10 Permits," May 16, 2024, 5, https://media.defense.gov/2024/May/16/2003467336/-1/-1/1/NWW%20RD%20SECTION%20404%20AND%20SECTION%2010-PERMITTING%20REFERENCE%20GUIDE%20(003).PDF; U.S. Army Corps of Engineers, "U.S. Army Corps of Engineers - Civil Works Tribal Consultation Policy," December 6, 2023, https://api.army.mil/e2/c/downloads/2023/12/06/f10ab368/dec2023-usace-tribal-consultation-policy.pdf; Advisory Council on Historic Preservation, "Consultation with Indian Tribes in the Section 106 Review Process: The Handbook," June 2021, https://www.achp.gov/sites/default/files/2021-06/ConsultationwithIndianTribesHandbook6-11-21Final.pdf; and Walla Walla District, U.S. Army Corps of Engineers, "Reference Guide for U.S. Army Corps of Engineers (Corps) Clean Water Act Section 404 & Rivers and Harbors Act Section 10 Permits," May 16, 2024, 40-41, https://media.defense.gov/2024/May/16/2003467336/-1/-1/1/NWW%20RD%20SECTION%20404%20AND%20SECTION%2010-PERMITTING%20REFERENCE%20GUIDE%20(003).PDF.

349 Environmental Protection Agency, "Mining Issues: EPA Region 9 Regional Tribal Operations Committee Meeting," presentation, April 20, 2023, slide 6, https://www.epa.gov/system/files/documents/2023-04/r9-rtoc-presentation-mining-issues-breakout-spring-2023.pdf.

350 Michael Bennon and Devon Wilson, "NEPA Litigation over Large Energy and Transport Infrastructure Projects," *Environmental Law Reporter* (October 2, 2023): 10839, https://ssrn.com/abstract=4498938.

351 Environmental Protection Agency, "Mining Issues: EPA Region 9 Regional Tribal Operations Committee Meeting," presentation, April 20, 2023, slide 18, https://www.epa.gov/system/files/documents/2023-04/r9-rtoc-presentation-mining-issues-breakout-spring-2023.pdf; and Talon Metals Corp. and DRA Americas Inc.,"NI 43-101 Technical Report Preliminary Economic Assessment (PEA) #3 of the Tamarack North Project - Tamarack, Minnesota," G4932-RPT-01, January 8, 2021, 293, https://talonmetals.com/wp-content/uploads/2021/02/Talon-Tamarack-PEA3_2021.pdf.

352 Environmental Protection Agency, "Mining Issues: EPA Region 9 Regional Tribal Operations Committee Meeting," presentation, April 20, 2023, slide 9, https://www.epa.gov/system/files/documents/2023-04/r9-rtoc-presentation-mining-issues-breakout-spring-2023.pdf.

353 Charles P. Nicholson ed., *2021 Annual NEPA Report* (Seattle, WA: National Environmental Policy Act Working Group, National Association of Environmental Professionals, July 2022), 9, https://naep.memberclicks.net/assets/annual-report/NEPA_Annual_Report_2021.pdf.

354 Council on Environmental Quality, "National Environmental Policy Act Implementing Regulations Revisions Phase 2," *Federal Register* 89, no. 85 (May 1, 2024): 35560, https://www.govinfo.gov/content/pkg/FR-2024-05-01/pdf/2024-08792.pdf. The Fiscal Responsibility Act of 2023 also established a one-year timeline for environmental assessments, but the deadline can similarly be extended by the lead federal agency. For more information on the impact of the Fiscal Responsibility Act of 2023 on the NEPA process, see "National Environmental Policy Act Implementing Regulations: Bipartisan Permitting Reform Implementation Final Rule," Council on Environmental Quality, May 2024, https://ceq.doe.gov/docs/laws-regulations/BPRI_Final_Rule_Slides_May_2024.pdf.

355 Talon Metals Corp. and DRA Americas Inc., "NI 43-101 Technical Report Preliminary Economic Assessment (PEA) #3 of the Tamarack North Project - Tamarack, Minnesota," G4932-RPT-01, January 8, 2021, 48, https://talonmetals.com/wp-content/uploads/2021/02/Talon-Tamarack-PEA3_2021.pdf.

356 Talon Metals Corp. and DRA Americas Inc., "NI 43-101 Technical Report Preliminary Economic Assessment (PEA) #3 of the Tamarack North Project - Tamarack, Minnesota," G4932-RPT-01, January 8, 2021, 293, 295–298, https://talonmetals.com/wp-content/uploads/2021/02/Talon-Tamarack-PEA3_2021.pdf.

357 Environmental Protection Agency, "Mining Issues: EPA Region 9 Regional Tribal Operations Committee Meeting," presentation, April 20, 2023, slide 7, https://www.epa.gov/system/files/documents/2023-04/r9-rtoc-presentation-mining-issues-breakout-spring-2023.pdf. See Environmental Protection Agency, "Mining Issues: EPA Region 9 Regional Tribal Operations Committee Meeting," presentation, April 20, 2023, slide 6, https://www.epa.gov/system/files/documents/2023-04/r9-rtoc-presentation-mining-issues-breakout-spring-2023.pdf.

358 Interagency Working Group on Mining Laws, Regulations, and Permitting, *Recommendations to Improve Mining on Public Lands*, 37; Bureau of Land Management, "Surface Management Handbook," H-3809-1, September 17, 2012, https://www.blm.gov/sites/blm.gov/files/H-3809-1.pdf; "Surface Management of Locatable Minerals," Bureau of Land Management, accessed September 23, 2024, https://www.blm.gov/programs/energy-and-minerals/mining-and-minerals/locatable-materials/surface-management; and *An Act to Establish Public Land Policy; to Establish Guidelines for its Administration; to Provide for the Management, Protection, Development, and Enhancement of the Public Lands; and for Other Purposes*, Public Law 94-579, *U.S. Statutes at Large* 90 (October 21, 1976): 2763, https://www.govinfo.gov/content/pkg/STATUTE-90/pdf/STATUTE-90-Pg2743.pdf. For regulations regarding unnecessary or undue degradation, see "Unnecessary or Undue Degradation," *Code of Federal Regulations*, Title 43 (1996): 3715.0-5, https://www.ecfr.gov/current/title-43/part-3710/subpart-3715#p-3715.0-5(Unnecessary%20or%20undue%20degradation); "Undue and Unnecessary Degradation," *Code of Federal Regulations*, Title 43 (2024): 3802.0-5(l), https://www.ecfr.gov/current/title-43/part-3800/section-3802.0-5#p-3802.0-5(l); and "How Do I Prevent Unnecessary or Undue Degradation While Conducting Operations on Public Lands?" *Code of Federal Regulations*, Title 43 (2001): 3809.415, https://www.ecfr.gov/current/title-43/subtitle-B/chapter-II/subchapter-C/part-3800/subpart-3809/subject-group-ECFR398287ff4adc2a4/section-3809.415.

359 Interagency Working Group on Mining Laws, Regulations, and Permitting, *Recommendations to Improve Mining on Public Lands*, 45, 87.

360 U.S. Government Accountability Office (GAO), *Hardrock Mining: BLM and Forest Service Have Taken Some Actions to Expedite the Mine Plan Review Process but Could Do More*, GAO-16-165 (Washington, DC: GAO, January 2016), 10–11, https://www.gao.gov/assets/gao-16-165.pdf.

361 Nicholson, *2021 Annual NEPA Repor*t, 9.

362 Interagency Working Group on Mining Laws, Regulations, and Permitting, *Recommendations to Improve Mining on Public Lands*, 53. For all types of projects in 2021, the average time from the BLM publishing a notice of intent for drafting an EIS to publishing a final EIS was 2.6 years. See Nicholson ed., 2021 *Annual NEPA Report*, 9.

363 Humphries, *Mining on Federal Lands*, 1, 3.

364 Carol Hardy Vincent and Erin H. Ward, *Withdrawal of Federal Lands: Analysis of a Common Legislated Withdrawal Provision*, CRS Report No. R46657 (Washington, DC: Congressional Research Service, January 2021), 7, https://crsreports.congress.gov/product/pdf/R/R46657; andBureau of Land Management, "Fluid Mineral Leases and Leasing Process," *Federal Register* 89, no. 79 (April 23, 2024): 30963, https://www.govinfo.gov/content/pkg/FR-2024-04-23/pdf/2024-08138.pdf.

365 GAO, *Federal Land Management*, 9–10.

366 Ibid., 15–16; Bureau of Land Management, "Required Fees for Mining Claims or Sites," *Federal Register* 89, no. 126 (July 1, 2024): 54365, https://www.govinfo.gov/content/pkg/FR-2024-07-01/pdf/2024-14301.pdf; and "Mining Claim Fees," Bureau of Land Management, accessed September 23, 2024, https://www.blm.gov/programs/energy-and-minerals/mining-and-minerals/locatable-minerals/mining-claims/fees.

367 GAO, Federal Land Management, 16; Interagency Working Group on Mining Laws, Regulations, and Permitting, *Recommendations to Improve Mining on Public Lands*, 9.

368 GAO, *Federal Land Management*, 9–10.

369 Ibid., 15–16.

370 Lithium Americas Corp. and M3 Engineering & Technology Corporation, "Feasibility Study, National Instrument 43-101 Technical Report for the Thacker Pass Project, Humboldt County, Nevada, USA," November 2, 2022, 34, 36, https://

s203.q4cdn.com/835901927/files/doc_financials/2023/ar/New-LAC-ThackerPassFeasibilityStudyNI43-101-October2023.pdf.

371 Lithium Americas Corp. and M3 Engineering & Technology Corporation, "Feasibility Study, National Instrument 43-101 Technical Report for the Thacker Pass Project, Humboldt County, Nevada, USA," November 2, 2022, 223, 225, https://s203.q4cdn.com/835901927/files/doc_financials/2023/ar/New-LAC-ThackerPassFeasibilityStudyNI43-101-October2023.pdf.

372 Winnemucca District Office, U.S. Bureau of Land Management, "Thacker Pass Lithium Mine Project: Record of Decision and Plan of Operations Approval," DOI-BLM-NV-W010-2020-0012-EIS, January 2021, 1, https://eplanning.blm.gov/public_projects/1503166/200352542/20033308/250039507/Thacker_Pass_Project_ROD_signed_2021-01-15.pdf; and Tim Burmeister, "Court Upholds Thacker Pass Approval," *Elko Daily*, March 6, 2024, https://elkodaily.com/news/local/business/mining/court-upholds-thacker-pass-approval/article_0ed30276-ccac-11ee-9e1a-6f91e6f45524.html.

373 Arnab Datta and James W. Coleman, "We Must End the Litigation Doom Loop," Slow Boring, April 28, 2024, https://www.slowboring.com/p/we-must-end-the-litigation-doom-loop. For policy ideas on addressing lengthy NEPA-related litigation, see Aidan Mackenzie, "Time Limit on Injunctions," Institute for Progress, April 28, 2023, https://ifp.org/time-limit-on-injunctions/.

374 Jervois and M3 Engineering & Technology Corporation, "Idaho Cobalt Operations: Form 43-101F1 Technical Report - Feasibility Study, Idaho, USA," M3-PN190348, November 13, 2020, 288, https://jervoisglobal.com/wp-content/uploads/2021/06/190348_Idaho_Cobalt_13112020_NI_43_101_Technical_Report-FILED-r1.pdf.

375 Jervois and M3 Engineering & Technology Corporation, "Idaho Cobalt Operations: Form 43-101F1 Technical Report - Feasibility Study, Idaho, USA," M3-PN190348, November 13, 2020, 288, https://jervoisglobal.com/wp-content/uploads/2021/06/190348_Idaho_Cobalt_13112020_NI_43_101_Technical_Report-FILED-r1.pdf.

376 U.S. Forest Service, "Idaho Cobalt Project Plan of Operations, Salmon-Challis National Forest, Lemhi County, ID," *Federal Register* 71, no. 211 (November 1, 2006): 64237-64240, https://www.govinfo.gov/content/pkg/FR-2006-11-01/pdf/E6-18362.pdf; and Jervois and M3 Engineering & Technology Corporation, "Idaho Cobalt Operations: Form 43-101F1 Technical Report - Feasibility Study, Idaho, USA," M3-PN190348, November 13, 2020, 50, https://jervoisglobal.com/wp-content/uploads/2021/06/190348_Idaho_Cobalt_13112020_NI_43_101_Technical_Report-FILED-r1.pdf; U.S. Forest Service, "Idaho Cobalt Project Plan of Operations, Salmon-Challis National Forest, Lemhi County, ID," *Federal Register* 66, no. 175 (September 10, 2001): 46992, https://www.govinfo.gov/content/pkg/FR-2001-09-10/pdf/01-22597.pdf; and Salmon-Cobalt Ranger District, Salmon-Challis National Forest, U.S. Forest Service, "Record of Decision: Idaho Cobalt Project," January 2009, https://www.fs.usda.gov/Internet/FSE_DOCUMENTS/fseprd640533.pdf.

377 Jervois and M3 Engineering & Technology Corporation, "Idaho Cobalt Operations: Form 43-101F1 Technical Report - Feasibility Study, Idaho, USA," M3-PN190348, November 13, 2020, 296, https://jervoisglobal.com/wp-content/uploads/2021/06/190348_Idaho_Cobalt_13112020_NI_43_101_Technical_Report-FILED-r1.pdf.

378 Ibid.

379 Federal Permitting Improvement Steering Council, "Adding Mining as a Sector of Projects Eligible for Coverage under Title 41 of the Fixing America's Surface Transportation Act," *Federal Register* 86, no. 5 (January 8, 2021): 1281-1288, https://www.govinfo.gov/content/pkg/FR-2021-01-08/pdf/2021-00088.pdf.

380 Federal Permitting Improvement Steering Council, "Adding Mining as a Sector of Projects Eligible for Coverage under Title 41 of the Fixing America's Surface Transportation Act," *Federal Register* 86, no. 5 (January 8, 2021): 1281-1288, https://www.govinfo.gov/content/pkg/FR-2021-01-08/pdf/2021-00088.pdf. As of September 2024, the Permitting Council was considering limiting coverage of projects in the mining sector to projects mining critical minerals and expanding coverage to projects beneficiating, processing, and recycling critical minerals. See Federal Permitting Improvement Steering Council, "Revising Scope of the Mining Sector of Projects That Are Eligible for Coverage under Title 41 of the Fixing America's Surface Transportation Act," *Federal Register* 88, no. 183 (September 22, 2023): 65350-65356, https://www.govinfo.gov/content/pkg/FR-2023-09-22/pdf/2023-20270.pdf.

381 "The Federal Permitting Improvement Steering Council (Permitting Council)," Federal Permitting Improvement Steering Council, September 9, 2022, 2, https://www.permits.performance.gov/sites/permits.dot.gov/files/2022-09/FPISC_090922.pdf.

382 "Hermosa Confirmed as the First FAST-41 Mining Project," South32 Limited, May 9, 2023, https://www.south32.net/news-media/latest-news/hermosa-confirmed-as-the-first-fast-41-mining-project. For the Hermosa project's projected permitting timeline, see "South32 Hermosa Critical Minerals Project," Permitting Dashboard: Federal Infrastructure Projects, accessed September 23, 2024, https://www.permits.performance.gov/permitting-project/fast-41-covered-projects/south32-hermosa-critical-minerals-project.

383 "Final Investment Approval to Develop Hermosa's Taylor Deposit," South32 Limited, February 15, 2024, https://www.south32.net/docs/default-source/exchange-releases/final-investment-approval-to-develop-hermosa-taylor-deposit-0x5ffd9fac3b216589.pdf?sfvrsn=5638590a_0; and South32 Hermosa Inc. and SWCA Environmental Consultants, "South32 Hermosa Inc.: Critical Minerals Exploration and Mine Plan of Operations," revised on December 1, 2023, prepared for Sierra Vista Ranger District, Coronado National Forest, 1-6, 2-1, 2-87, 3-3, https://south32hermosa.com/wp-content/uploads/2024/01/2024_01_South32-Hermosa-Critical-Minerals-Plan-of-Operations_RE.pdf.

384 "Final Investment Approval to Develop Hermosa's Taylor Deposit," South32 Limited, February 15, 2024, 12, https://www.

south32.net/docs/default-source/exchange-releases/final-investment-approval-to-develop-hermosa-taylor-deposit-0x5ffd-9fac3b216589.pdf?sfvrsn=5638590a_0.

385 Displaying support for environmentally responsible permitting, President Donald J. Trump's Executive Order 13817 said, "It shall be the policy of the Federal Government to reduce the Nation's vulnerability to disruptions in the supply of critical minerals, which constitutes a strategic vulnerability for the security and prosperity of the United States. The United States will further this policy for the benefit of the American people and in a safe and environmentally responsible manner." Similarly, the Biden-Harris administration declared, "To meet current and future demand, and to break our reliance on single sources while creating good jobs for American workers, mining reform should assure that a reliable and sustainable supply of critical minerals can be provided both through environmentally and socially responsible mining and processing projects and other sustainable sources." See "Executive Order 13817 of December 20, 2017: A Federal Strategy To Ensure Secure and Reliable Supplies of Critical Minerals," *Federal Register* 82, no. 246 (December 26, 2017): 60835-60837, https://www.govinfo.gov/content/pkg/FR-2017-12-26/pdf/2017-27899.pdf; and "Biden-Harris Administration Fundamental Principles for Domestic Mining Reform," U.S. Department of the Interior, February 22, 2022, 1, https://www.doi.gov/sites/doi.gov/files/biden-harris-administration-fundamental-principles-for-domestic-mining-reform.pdf.

386 "Executive Order 13953 of September 30, 2020: Addressing the Threat to the Domestic Supply Chain from Reliance on Critical Minerals from Foreign Adversaries and Supporting the Domestic Mining and Processing Industries," *Federal Register* 85, no. 193 (October 5, 2020): 62543, https://www.govinfo.gov/content/pkg/FR-2020-10-05/pdf/2020-22064.pdf.

387 U.S. Congress, Senate, *Energy Permitting Reform Act of 2024*, S 4573, 118th Cong., 2d sess., introduced in Senate July 23, 2024, https://www.congress.gov/bill/118th-congress/senate-bill/4753/text.

388 Under federal regulations, "Consultation means the process of seeking, discussing, and considering the views of other participants, and, where feasible, seeking agreement with them regarding matters arising in the section 106 process." See "Definitions," *Code of Federal Regulations*, Title 36 (2004), §§ 800.16(f), https://www.ecfr.gov/current/title-36/part-800/section-800.16#p-800.16(f).

389 Samuel Block, "Mining Energy-Transition Metals: National Aims, Local Conflicts," MSCI, June 3, 2021, https://www.msci.com/www/blog-posts/mining-energy-transition-metals/02531033947.

390 GAO, *Federal Land Management*, 34-35.

391 A future administration could adopt rules that establish more rigorous standards for Tribal consultation, such as those described in H.R. 3587. See U.S. Congress, House, *Requirements, Expectations, and Standard Procedures for Effective Consultation with Tribes Act (RESPECT Act)*, HR 3587, 117th U.S. Cong., 1st sess., introduced in House May 28, 2021, https://www.congress.gov/117/bills/hr3587/BILLS-117hr3587ih.pdf.

392 Federal Permitting Improvement Steering Council, "FAST-41 and Permitting Council," February 2022, slide 9, https://www.nga.org/wp-content/uploads/2022/03/Permitting-Council-and-FAST-41-Overview_2022.pdf; "Does My Project Qualify as a FAST-41 Covered Project?," Federal Permitting Improvement Steering Council, October 2019, https://www.permits.performance.gov/sites/permits.dot.gov/files/2019-10/Flowchart%20for%20Covered%20Project%20Definition.pdf; and Department of Energy, "Notice of Final Determination on 2023 Critical Materials List," *Federal Register* 88, no. 149 (August 4, 2023): 51792, https://www.govinfo.gov/content/pkg/FR-2023-08-04/pdf/2023-16611.pdf. The Permitting Council can be heavily influenced by the president because the council includes a presidentially appointed executive director, the Office of Management and Budget director, the chairperson of the CEQ, and deputy secretaries or equivalent designees from 13 agencies. See "Permitting Council," Permitting Dashboard: Federal Infrastructure Projects, accessed September 23, 2024, https://www.permits.performance.gov/fpisc-content/permitting-council-0. The Department of Energy's list of minerals deemed "critical materials for energy" include aluminum, cobalt, copper, dysprosium, gallium, iridium, lithium, magnesium, natural graphite, neodymium, nickel, platinum, praseodymium, terbium, and silicon. See Department of Energy, "Notice of Final Determination on 2023 Critical Materials List," *Federal Register* 88, no. 149 (August 4, 2023): 51792, https://www.govinfo.gov/content/pkg/FR-2023-08-04/pdf/2023-16611.pdf.

393 "Permitting Council Announces First-Ever Critical Minerals Mining Project to Gain FAST-41 Coverage," Permitting Council Press Office, May 9, 2023, https://www.permits.performance.gov/fpisc-content/permitting-council-announces-first-ever-critical-minerals-mining-project-gain-fast-41; and Federal Permitting Improvement Steering Council, "Adding Mining as a Sector of Projects Eligible for Coverage under Title 41 of the Fixing America's Surface Transportation Act," *Federal Register* 86, no. 5 (January 8, 2021): 1281-1288, https://www.govinfo.gov/content/pkg/FR-2021-01-08/pdf/2021-00088.pdf.

394 "The Federal Permitting Improvement Steering Council (Permitting Council)," Federal Permitting Improvement Steering Council, September 9, 2022, https://www.permits.performance.gov/sites/permits.dot.gov/files/2022-09/FPISC_090922.pdf; and "Does My Project Qualify as a FAST-41 Covered Project?," Federal Permitting Improvement Steering Council, October 2019, https://www.permits.performance.gov/sites/permits.dot.gov/files/2019-10/Flowchart%20for%20Covered%20Project%20Definition.pdf. The Idaho Cobalt Operations' total estimated capital costs for the life of the mine is $136 million. See Jervois and M3 Engineering & Technology Corporation, "Idaho Cobalt Operations: Form 43-101F1 Technical Report - Feasibility Study, Idaho, USA," M3-PN190348, November 13, 2020, 8, https://jervoisglobal.com/wp-content/

uploads/2021/06/190348_Idaho_Cobalt_13112020_NI_43_101_Technical_Report-FILED-r1.pdf.

395 A study assessing U.S. Forest Service data from 2005 to 2020 found that the average time to complete an environmental assessment was 1.7 years, while the average time to complete an environmental impact statement was 3.4 years. See John Ruple et al., "Evidence-Based Recommendations for Improving National Environmental Policy Act Implementation," *Columbia Journal of Environmental Law* 47 (Spring 2022). 293, https://journals.library.columbia.edu/index.php/cjel/article/view/9479/6000; *An Act to Establish a National Policy for the Environment, to Provide for the Establishment of a Council on Environmental Quality, and for Other Purposes*, Pub. L. 91-190, *U.S. Statutes at Large* 83 (January 1, 1970): 853, https://www.govinfo.gov/content/pkg/STATUTE-83/pdf/STATUTE-83-Pg852.pdf#page=1.

396 Council on Environmental Quality, "National Environmental Policy Act Implementing Regulations Revisions Phase 2," *Federal Register* 89, no. 85 (May 1, 2024): 35557, https://www.govinfo.gov/content/pkg/FR-2024-05-01/pdf/2024-08792.pdf; and "National Environmental Policy Act: NEPA Overview," U.S. Army Corps of Engineers, January 2019, 2, https://usace.contentdm.oclc.org/digital/api/collection/p16021coll11/id/3378/download.

397 As the Ninth Circuit Court of Appeals wrote in *LaFlamme v. F.E.R.C.* (1988), "The plaintiff need not show that significant effects *will in fact occur*, but if the plaintiff raises substantial questions whether a project may have a significant effect, an [environmental impact statement] *must* be prepared." The Ninth Circuit Court of Appeals has ruled and reaffirmed in multiple cases that only one factor has to trigger for an environmental impact statement to be necessary. See *LaFlamme v. F.E.R.C.*, 852 F.2d 389 (9th Cir. 1988), 397, https://casetext.com/case/laflamme-v-ferc#p397; *National Parks Conservation Ass'n v. Babbitt*, 241 F.3d 722 (9th Cir. 2001), 730-732, https://casetext.com/case/national-parks-conservation-assn-v-babbitt; *Ocean Advocates v. U.S. Army Corps of Engineers*, 402 F.3d 846 (9th Cir. 2004), 865, https://casetext.com/case/ocean-advocates-v-us-army-corps-of-engineers-2; and *Environmental Defense Center v. Bureau of Ocean Energy Management*, 36 F.4th 850 (9th Cir. 2022), 878-879, https://casetext.com/case/envtl-def-ctr-v-bureau-of-ocean-energy-mgmt; and "Regulatory Program Preparing and Coordinating Environmental Impact Statements," South Pacific Division, U.S. Army Corps of Engineers, February 8, 2013, https://www.spd.usace.army.mil/Portals/13/docs/regulatory/qmsref/eis/12509.pdf.

398 GAO, *Federal Land Management*, 40-41.

399 GAO, *Hardrock Mining*, 22.

400 Ibid., 22-25; and GAO, Federal Land Management, 40-41.

401 GAO, Hardrock Mining, 22; and GAO, *Federal Land Management*, 40-41.

402 Other governments have provided financial support—such as funds to purchase new equipment—to private industry in their countries when imposing high environmental standards, helping offset costs associated with compliance. See Comptroller General of the United States, *The U.S. Mining and Mineral-Processing Industry*, 28.

403 U.S. Department of Defense, "DoD Issues $24.8M Critical Minerals Award to Perpetua Resources," press release, December 19, 2022, https://www.defense.gov/News/Releases/Release/Article/3249350/dod-issues-248m-critical-minerals-award-to-perpetua-resources/; and U.S. Department of Defense, "Department of Defense Awards $14.7 Million to Enhance North American Cobalt and Graphite Supply Chain," press release, May 16, 2024, https://www.defense.gov/News/Releases/Release/Article/3777044/.

404 GAO, *Hardrock Mining*, 22, footnote a, 30.

405 U.S. Congress, Senate, *Building Chips in America Act of 2023*, S 2228, 118th Cong., 1st sess., passed House September 23, 2024, https://www.congress.gov/118/bills/s2228/BILLS-118s2228es.pdf; and "Bill Signed: S. 2228," The White House, October 2, 2024, https://www.whitehouse.gov/briefing-room/legislation/2024/10/02/bill-signed-s-2228.

406 John Ruple et al., "Evidence-Based Recommendations for Improving National Environmental Policy Act Implementation," *Columbia Journal of Environmental Law* 47 (Spring 2022): 293, https://ssrn.com/abstract=3965925.

407 Department of Energy, "Notice of Final Determination on 2023 Critical Materials List," *Federal Register* 88, no. 149 (August 4, 2023): 51792, https://www.govinfo.gov/content/pkg/FR-2023-08-04/pdf/2023-16611.pdf.

408 George Otis Smith, "The Economic Limits to Domestic Independence in Minerals," in *Mineral Resources of the United States 1917*, Part 1, ed. H. D. McCaskey (Washington, DC: U.S. Government Printing Office, 1921), 6a, https://babel.hathitrust.org/cgi/pt?id=uc1.$b570206&seq=20.

409 National Minerals Information Center, "Mineral Commodity Summaries 2024: RARE EARTHS Data Release," U.S. Geological Survey, January 30, 2024, https://data.usgs.gov/datacatalog/data/USGS:65b7d85dd34e36a39045b50b.

410 National Minerals Information Center, "Mineral Commodity Summaries 2024: COPPER Data Release," U.S. Geological Survey, January 30, 2024, https://data.usgs.gov/datacatalog/data/USGS:65b7d77ed34e36a39045b4b2.

411 Matthew Blackwood and Catherine DeFilippo, *Germanium and Gallium: U.S. Trade and Chinese Export Controls* (Washington, DC: U.S. International Trade Commission, March 2024), 1, https://www.usitc.gov/publications/332/executive_briefings/ebot_germanium_and_gallium.pdf.

412 International Energy Agency (IEA), *Global Critical Minerals Outlook 2024* (Paris: IEA, May 2024), 182, https://www.iea.org/reports/global-critical-minerals-outlook-2024.

413 Siyi Liu and Dominique Patton, "China bans export of rare earths processing tech over national security," Reuters, December 22, 2023, https://www.reuters.com/markets/commodities/china-bans-export-rare-earths-processing-technologies-2023-12-21/.

414 IEA, *Critical Minerals Market Review 2023* (Paris, France: IEA, July 2023), 68, https://www.iea.org/reports/critical-minerals-market-review-2023.

415 Timur Abenov et al., *Has mining lost its luster? Why talent is moving elsewhere–and how to bring them back* (New York: McKinsey & Company, June 2023), https://www.mckinsey.com/industries/metals-and-mining/our-insights/has-mining-lost-its-luster-why-talent-is-moving-elsewhere-and-how-to-bring-them-back.

416 U.S. House of Representatives, *Select Committee on the Strategic Competition Between the United States and the Chinese Communist Party*, Committee Report 118-3 (Washington, DC: U.S. Congress, December 2023), https://docs.house.gov/meetings/ZS/ZS00/20231212/116682/HRPT-118-3.pdf.

417 "Revising Scope of the Mining Sector of Projects That Are Eligible for Coverage Under Title 41 of the Fixing America's Surface Transportation Act," *Federal Register*, September 22, 2023, https://www.federalregister.gov/documents/2023/09/22/2023-20270/revising-scope-of-the-mining-sector-of-projects-that-are-eligible-for-coverage-under-title-41-of-the.

418 IEA, "The Role of Critical Minerals in Clean Energy Transitions: Executive Summary." (Paris: IEA, March 2022), https://www.iea.org/reports/the-role-of-critical-minerals-in-clean-energy-transitions/executive-summary.

419 Katherine Zimmerman, "Why the US Should Spend 0.3 Percent of Its Defense Budget to Prevent an African Debacle," *Military Times*, March 12, 2020, https://www.militarytimes.com/opinion/commentary/2020/03/12/why-the-us-should-spend-03-percent-of-its-defense-budget-to-prevent-an-african-debacle/.

420 Sara Harcourt and Jorge Rivera, "Official Development Assistance (ODA)," One Campaign, accessed December 1, 2024, https://data.one.org/topics/official-development-assistance/; and The White House, *U.S. Strategy Toward Sub-Saharan Africa* (Washington, DC: The White House, August 2022), https://www.whitehouse.gov/wp-content/uploads/2022/08/U.S.-Strategy-Toward-Sub-Saharan-Africa-FINAL.pdf.

421 Peter J. Meyer, *U.S. Foreign Assistance to Latin America and the Caribbean: FY2024 Appropriations*, CRS Report No. R47721 (Washington, DC: Congressional Research Service, updated August 6, 2024), https://sgp.fas.org/crs/row/R47721.pdf.

422 Office of the Spokesperson, U.S. Department of State, "Partnership for Global Infrastructure and Investment Lobito Corridor: Supporting Transcontinental Connectivity," press release, September 24, 2024, https://www.state.gov/partnership-for-global-infrastructure-and-investment-lobito-corridor-supporting-transcontinental-connectivity/; and Office of the Spokesperson, U.S. Department of State, "United States Welcomes New Investment in Rare Earth Element Production for Serra Verde Project in Brazil," press release, October 21, 2024, https://www.state.gov/united-states-welcomes-new-investment-in-rare-earth-element-production-for-serra-verde-project-in-brazil/.

423 "Nickel Facts," Natural Resources Canada, Government of Canada, modified February 15, 2024, https://natural-resources.canada.ca/our-natural-resources/minerals-mining/mining-data-statistics-and-analysis/minerals-metals-facts/nickel-facts/20519; and Eri Silva, "Indonesian Nickel Production Dominates Commodity Market," S&P Global, February 6, 2024, https://www.spglobal.com/marketintelligence/en/news-insights/latest-news-headlines/indonesian-nickel-production-dominates-commodity-market-80242322.

424 "Metals & Mining. S&P Global Market Intelligence, Capital IQ Pro (database), accessed December 10, 2024, https://www.capitaliq.spglobal.com/web/client?auth=inherit#dashboard/metalsAndMining.

425 The White House, "Joint Statement from the Leaders of the United States and the Republic of Indonesia: Elevating Relations to a Comprehensive Strategic Partnership," press release, November 13, 2023, https://www.whitehouse.gov/briefing-room/statements-releases/2023/11/13/joint-statement-from-the-leaders-of-the-united-states-and-the-republic-of-indonesia-elevating-relations-to-a-comprehensive-strategic-partnership/; Valdya Baraputri, "The Rush for Nickel: 'They Are Destroying Our Future'," BBC, July 10, 2023, https://www.bbc.com/news/world-asia-66131451; and Climate Rights International, *Nickel Unearthed: The Human and Climate Costs of Indonesia's Nickel Industry* (Climate Rights International, January 2024), https://cri.org/wp-content/uploads/2024/01/NICKEL_UNEARTHED.pdf.

426 Marek Hanusch et al., *Digging Beneath the Surface: An Exploration of the Net Benefits of Mining in Southern Africa* (Washington, DC: World Bank, June 2019), https://documents1.worldbank.org/curated/ru/506751562777260359/pdf/Digging-Beneath-the-Surface-An-Exploration-of-the-Net-Benefits-of-Mining-in-Southern-Africa.pdf.

427 "Democratic Republic of the Congo - Mining and Minerals," International Trade Administration, last published date March 14, 2024, https://www.trade.gov/country-commercial-guides/democratic-republic-congo-mining-and-minerals; "Cobalt Facts," Natural Resources Canada, Government of Canada, modified November 7, 2023, https://natural-resources.canada.ca/our-natural-resources/minerals-mining/mining-data-statistics-and-analysis/minerals-metals-facts/cobalt-facts/24981; and "Copper Facts," Natural Resources Canada, Government of Canada, modified July 26, 2024, https://natural-resources.canada.ca/our-natural-resources/minerals-mining/mining-data-statistics-and-analysis/minerals-metals-facts/copper-facts/20506.

428 Mitra Taj and Susan Taylor, "Freeport Closes Congo Mine Sale; Trump Appointees Keenly Awaited," Reuters, November 17, 2016, https://www.reuters.com/article/us-freeport-mcmoran-mine-ceo-idUSKBN13B2SU/; and Andrew Topf, "First Quantum Gets out of Congo with $1.25 Billion Asset Sale," Mining.com, January 5, 2012, https://www.mining.com/first-quantum-gets-out-of-congo-with-1-25-billion-asset-sale/.

429 Topf, "First Quantum Gets out of Congo."

430 Vivek Salgaocar, "Why Innovation in the Mining Sector Is Critical for the Energy Transition," World Economic Forum, November 2, 2022, https://www.weforum.org/stories/2022/11/why-innovation-in-the-mining-sector-is-critical-for-the-energy-transition/.

431 Emmanuel Aramendia et al., "Global Energy Consumption of the Mineral Mining Industry: Exploring the Historical Perspective and Future Pathways to 2060," *Global Environmental Change* 83, 102745 (December 2023), https://doi.org/10.1016/j.gloenvcha.2023.102745.

432 "Saudi Arabia Sets New World Record in Producing Low-Cost Electricity from Wind Energy," KAPSARC, May 22, 2024, https://www.kapsarc.org/news/saudi-arabia-sets-new-world-record-in-producing-low-cost-electricity-from-wind-energy/.

433 GBO Correspondent, "With Record Low Cost, Saudi Arabia Leads Middle East's Solar Revolution," *Global Business Outlook*, June 5, 2024, https://globalbusinessoutlook.com/energy/with-record-low-cost-saudi-arabia-leads-middle-easts-solar-revolution/.

434 "Access to Electricity (% of Population)," Data, World Bank Group, accessed December 1, 2024, https://data.worldbank.org/indicator/EG.ELC.ACCS.ZS?most_recent_value_desc=false.

435 Anglo American, "Anglo American Partners with EDF Renewables to Secure 100% Renewable Energy Supply for South Africa Operations," press release, March 18, 2022, https://www.angloamerican.com/media/press-releases/2022/18-03-2022.

436 "The Top 50 Container Ports," World Shipping Council, accessed December 1, 2024, https://www.worldshipping.org/top-50-ports.

437 "Transport Efficiency," Mineral Products Association, accessed December 1, 2024, https://www.mineralproducts.org/Industry-Overview/Sustainable-Solutions/Transport-Efficency.aspx.

438 "Africa's Surprising New Age of Rail," *The Economist*, July 18, 2024, https://www.economist.com/middle-east-and-africa/2024/07/18/africas-surprising-new-age-of-rail.

439 Ibid.

440 "Connecting the Democratic Republic of the Congo, Zambia, and Angola to Global Markets through the Lobito Corridor," International Partnerships, European Commission, accessed December 1, 2024, https://international-partnerships.ec.europa.eu/policies/global-gateway/connecting-democratic-republic-congo-zambia-and-angola-global-markets-through-lobito-corridor_en.

441 Vincent Rouget, "From Licence to Operate to Social Impact: What 'Good Mining' Looks like Is Evolving," Control Risks, April 28, 2022, https://www.controlrisks.com/our-thinking/insights/from-licence-to-operate-to-social-impact?utm_referrer=https%3A%2F%2Fwww.google.com.

442 Cecilia Jamasmie, "Rise of Ethical Investing, Social License the Top Risks to Miners," Mining.com, October 2, 2019, https://www.mining.com/stakeholder-pressure-ethical-investing-make-losing-licence-to-operate-biggest-threat-to-miners/.

443 Paul Mitchell, "Top 10 Risks and Opportunities for Mining and Metals Companies in 2025," EY, October 1, 2024, https://www.ey.com/en_gl/insights/energy-resources/risks-opportunities.

444 Author's analysis using data from S&P Global Capital IQ Pro at https://www.capitaliq.spglobal.com/.

445 Andres Schipani, "Protests Threaten to Dent the Outlook for Peruvian Copper," *Financial Times*, October 19, 2023, https://www.ft.com/content/a9f3e962-9465-45ea-94a6-ccc245ba0ad9.

446 "Social Unrest an Operational Risk to Some Peru-Focused Mining Issuers," Fitch Ratings, March 9, 2023, https://www.fitchratings.com/research/corporate-finance/social-unrest-operational-risk-to-some-peru-focused-mining-issuers-09-03-2023.

447 "Panama - Country Commercial Guide - Market Challenges," (Washington, DC: International Trade Administration, April 2023), https://www.trade.gov/country-commercial-guides/panama-market-challenges; "2024 Investment Climate Statements: Panama," U.S. Department of State, accessed December 30, 2024, https://www.state.gov/reports/2024-investment-climate-statements/panama/.

448 Jacob Lorinc, "Cobre Panama: How a $10 Billion Copper Mine Is Now Sitting Idle in the Jungle," Mining.com, April 16, 2024, https://www.mining.com/web/cobre-panama-how-a-10-billion-copper-mine-is-now-sitting-idle-in-the-jungle; Cristina Guevara, "Panama's Mining Future Is at a Tipping Point," *Foreign Policy*, November 21, 2023, https://foreignpolicy.com/2023/11/21/panama-mining-protests-contract-cortizo-fqm-mineral/; and Damien Nyer, "Panama's Reckless Gamble with Foreign Investments," White & Case, July 19, 2024, https://www.whitecase.com/insight-our-thinking/panamas-reckless-gamble-foreign-investments.

449 Anastasia Lyrchikova, "Russia Says U.S. Ban on Russian Nuclear Fuel Will Harm Global Market," Reuters, May 14, 2024, https://www.reuters.com/business/energy/russia-says-us-ban-russian-nuclear-fuel-will-harm-global-market-2024-05-14/; Gracelin Baskaran, "Seven Recommendations for the New Administration and Congress: Building U.S. Critical Minerals Security," CSIS, *Commentary*, November 14, 2024, https://www.csis.org/analysis/seven-recommendations-new-administration-and-congress-building-us-critical-minerals; Harry Dempsey and Joseph Cotterill, "How China Is Winning the Race for Africa's Lithium," *Financial Times*, April 3, 2023, https://www.ft.com/content/02d6f35d-e646-40f7-894c-ffc-c6acd9b25; "Namibia - Country Commercial Guide - Mining and Minerals," International Trade Administration, February 29, 2024, https://www.trade.gov/country-commercial-guides/namibia-mining-and-minerals; and Cy McGeady and Gracelin Baskaran, "Resource Nationalism Is Not the United States' Biggest Minerals Problem," CSIS, *Commentary*, September 15, 2023, https://www.csis.org/analysis/resource-nationalism-not-united-states-biggest-minerals-problem.

450 Eri Silva, "Indonesian Nickel Production Dominates Commodity Market," S&P Global, February 6, 2024, https://www.spglobal.com/marketintelligence/en/news-insights/

latest-news-headlines/indonesian-nickel-production-dominates-commodity-market-80242322.

451 BHP, "Western Australia Nickel to Temporarily Suspend Operations," press release, July 11, 2024, https://www.bhp.com/news/media-centre/releases/2024/07/western-australia-nickel-to-temporarily-suspend-operations; and Cecilia Jamasmie, "Glencore to Sell Stake in New Caledonia Nickel Operation," Mining.com, February 12, 2024, https://www.mining.com/glencore-to-sell-stake-in-troubled-new-caledonia-nickel-operation/.

452 Gracelin Baskaran et al., Latin America: *The World's Copper Stronghold* (Washington, DC: CSIS, November 13, 2024), https://www.csis.org/analysis/latin-america-worlds-copper-stronghold-0.

453 Divya Rajagopal, "First Quantum Minerals Jumps 5% on Panama Plans for Mine Environmental Audit," Reuters, July 2, 2024, https://www.reuters.com/markets/commodities/first-quantum-minerals-jumps-6-panama-plans-mine-environmental-audit-2024-07-02/.

454 James Cust and Albert Zeufack, eds., *Africa's Resource Future: Harnessing Natural Resources for Economic Transformation during the Low-Carbon Transition* (Paris: Agence française de développement; and Washington, DC: World Bank, 2023), https://openknowledge.worldbank.org/server/api/core/bitstreams/2cfa1ec3-1318-4415-aa8e-c8f3ce5e11ab/content; and "Zambia Plans to Import and Ration Electricity Due to Drought," Reuters, February 29, 2024, https://www.reuters.com/business/environment/zambia-plans-import-ration-electricity-due-drought-2024-02-29/.

455 Gracelin Baskaran, "Gracelin Baskaran: Zambia Shows It's Not Too Late for SA's Mining Sector," BusinessLIVE, February 16, 2023, https://www.businesslive.co.za/bd/opinion/columnists/2023-02-16-gracelin-baskaran-zambia-shows-its-not-too-late-for-sas-mining-sector/; and Gracelin Baskaran, "Partnering with Middle Eastern Countries to Boost U.S. Minerals Security," CSIS, *Commentary*, September 6, 2024, https://www.csis.org/analysis/partnering-middle-eastern-countries-boost-us-minerals-security.

456 Countries with free trade agreements and/or other trade agreements performed well. However, the Democratic Republic of the Congo (DRC) was color-coded red. Despite being a beneficiary of the African Growth and Opportunity Act, the DRC is also subject to Dodd-Frank sanctions. According to the U.S. Trade Representative, $149 million worth of U.S. goods were exported to the DRC in 2022 and $183 million in goods were imported—both figures represent a decrease from the previous year.

457 "Sourcing Critical Minerals to Support the Global Clean Energy Transition," U.S. International Development Finance Corporation, accessed December 1, 2024, https://www.dfc.gov/investment-story/sourcing-critical-minerals-support-global-clean-energy-transition.

458 Frik Els, "CHART: China's Belt and Road Mining Investment Hits Record," Mining.com, February 20, 2024. https://www.mining.com/chart-chinas-belt-and-road-mining-investment-hits-record.

459 Christoph Nedopil Wang, *China Belt and Road Initiative (BRI) Investment Report 2023* (Shanghai: Green Finance & Development Center, February 2024), https://greenfdc.org/china-belt-and-road-initiative-bri-investment-report-2023/.

460 Baskaran et al., Latin America: *The World's Copper Stronghold*.

461 "Power Africa 10th Anniversary," U.S. Agency for International Development, accessed December 1, 2024, https://www.usaid.gov/powerafrica/10th-anniversary.

462 "Could AI Help Find Valuable Mineral Deposits?," *The Economist*, November 1, 2023, https://www.economist.com/science-and-technology/2023/11/01/could-ai-help-find-valuable-mineral-deposits.

463 Sarah Way, "The Strategies Driving the Players in Competition for Africa's Critical Minerals," *Atlantic Council* (blog), September 9, 2024, https://www.atlanticcouncil.org/blogs/africasource/the-strategies-driving-the-players-in-competition-for-africas-critical-minerals/.

464 Eduardo Baptista et al., "Starting Latin America Trip, Xi Jinping Opens Huge Port in Peru Funded by China," Reuters, November 15, 2024, https://www.reuters.com/world/chinas-xi-arrives-lima-apec-open-pacific-megaport-2024-11-14/.

465 Diana Roy, "China's Growing Influence in Latin America," Council on Foreign Relations, June 15, 2023. https://www.cfr.org/backgrounder/china-influence-latin-america-argentina-brazil-venezuela-security-energy-bri.

466 Baskaran, "Seven Recommendations for the New Administration and Congress."

467 Norwegian Ministry of Energy, "Public Consultation of the First Licensing Round for Seabed Minerals," press release, June 26, 2024, https://www.regjeringen.no/en/aktuelt/public-consultation-of-the-first-licensing-round-for-seabed-minerals/id3047008/.

468 Japan Oil, Gas, and Metals National Corporation, "JOGMEC Conducts World's First Successful Excavation of Cobalt-Rich Seabed in the Deep Ocean; Excavation Test Seeks to Identify Best Practices to Access Essential Green Technology Ingredients While Minimizing Environmental Impact ," press release, August 21, 2020, https://www.jogmec.go.jp/english/news/release/news_01_000033.html; and Jiji Press, "Over 200 Million Metric Tons of Rare Metals Found near Remote Tokyo Island," *Japan Times*, June 22, 2024, https://www.japantimes.co.jp/news/2024/06/22/japan/science-health/tokyo-island-rare-metals-find/.

469 Nitin Kumar, "Centre Unveils Rules to Boost Offshore Mineral Exploration & Production," *Business Standard* (India), August 12, 2024, https://www.business-standard.com/industry/news/centre-unveils-offshore-areas-mineral-trust-rules-to-boost-exploration-124081201657_1.html.

470 Pete McKenzie, "The Tiny Nation at the Vanguard of Mining the Ocean Floor," *New York Times*, May 9, 2024, https://www.

nytimes.com/2024/05/09/world/asia/cook-islands-seabed-mining.html.

471 Evan Bloom, "Five Takeaways from the US Continental Shelf Announcement," Wilson Center, January 3, 2024, https://www.wilsoncenter.org/article/five-takeaways-us-continental-shelf-announcement.

472 James R. Hein et al., "Deep-Ocean Mineral Deposits as a Source of Critical Metals for High- and Green-Technology Applications: Comparison with Land-Based Resources," *Ore Geology Reviews* 51 (June 2013): 1-14, https://doi.org/10.1016/j.oregeorev.2012.12.001.

473 Andrea Koschinsky et al., "Deep Sea Mining: Interdisciplinary Research on Potential Environmental, Legal, Economic, and Societal Implications," *Integrated Environmental Assessment and Management* 14, no. 6 (November 2018): 672-91, https://doi.org/10.1002/ieam.4071.

474 James R. Hein and Kira Mizell, "Deep-Ocean Polymetallic Nodules and Cobalt-Rich Ferromanganese Crusts in the Global Ocean: New Sources for Critical Metals," in Alfonso Ascencio-Herrera and Myron H. Nordquist (eds.), *The United Nations Convention on the Law of the Sea, Part XI Regime and the International Seabed Authority: A Twenty-Five Year Journey* (Leiden: Brill Nijhoff, 2022), 177-97, https://doi.org/10.1163/9789004507388_013.

475 Koschinsky et al., "Deep Sea Mining."

476 "Exploration Contracts," International Seabed Authority, accessed September 18, 2024, https://www.isa.org.jm/exploration-contracts/.

477 The Metals Company, "NORI and Allseas Lift Over 3,000 Tonnes of Polymetallic Nodules to Surface from Planet's Largest Deposit of Battery Metals, as Leading Scientists and Marine Experts Continue Gathering Environmental Data," press release, November 14 2022, https://investors.metals.co/news-releases/news-release-details/nori-and-allseas-lift-over-3000-tonnes-polymetallic-nodules/; and DEME Group, "Deep-Seabed Mining Robot Patania II Successfully Reconnected - Mission Continues," press release, April 26, 2021, https://www.deme-group.com/news/deep-seabed-mining-robot-patania-ii-successfully-reconnected-mission-continues.

478 International Seabed Authority, "Beijing Pioneer Hi-Tech Development Corporation Ltd. Launches Stakeholder Consultation on Environmental Impact Statement for Polymetallic Nodule Mining Component Test," press release, May 1, 2024, https://www.isa.org.jm/news/beijing-pioneer-hi-tech-development-corporation-ltd-launches-stakeholder-consultation-on-environmental-impact-statement-for-polymetallic-nodule-mining-component-test/; and International Seabed Authority, "China Minmetals Corporation Launches Stakeholder Consultation on Environmental Impact Statement for Polymetallic Nodule Collector Test," press release, April 23, 2024, https://www.isa.org.jm/news/china-minmetals-corporation-launches-stakeholder-consultation-on-environmental-impact-statement-for-polymetallic-nodule-collector-test/.

479 Seaver Wang, "Sparing the Land by Collecting Minerals at Sea," Breakthrough Institute, February 14, 2024, https://thebreakthrough.org/issues/energy/sparing-the-land-by-collecting-minerals-at-sea.

480 "Infographic: China's Lithium Ion Battery Supply Chain Dominance," Benchmark Source, October 3, 2022, https://source.benchmarkminerals.com/article/infographic-chinas-lithium-ion-battery-supply-chain-dominance; and U.S. Geological Survey, *Mineral Commodity Summaries 2024* (Washington, DC: U.S. Geological Survey, 2024), 65, https://doi.org/10.3133/mcs2024.

481 International Energy Agency, *Global EV Outlook 2024: Moving Towards Increased Affordability* (Paris: International Energy Agency, 2024), 80, https://www.iea.org/reports/global-ev-outlook-2024.

482 Internal Revenue Service, "Clean Vehicle Credits under Sections 25E and 30D; Transfer of Credits; Critical Minerals and Battery Components; Foreign Entities of Concern," *Federal Register* 89, no. 88 (May 6, 2024): 37706, https://www.federalregister.gov/documents/2024/05/06/2024-09094/clean-vehicle-credits-under-sections-25e-and-30d-transfer-of-credits-critical-minerals-and-battery.

483 Eri Silva, "Deep-Sea Mining Still Profitable with Low Nickel Prices, Prospective Miners Say," S&P Global, March 25, 2024, https://www.spglobal.com/marketintelligence/en/news-insights/latest-news-headlines/deep-sea-mining-still-profitable-with-low-nickel-prices-prospective-miners-say-80892143; and Kris Van Nijen, Steven Van Passel, and Dale Squires, "A Stochastic Techno-economic Assessment of Seabed Mining of Polymetallic Nodules in the Clarion Clipperton Fracture Zone," *Marine Policy* 95 (September 2018): 133-41, https://doi.org/10.1016/j.marpol.2018.02.027.

484 Raúl M. Grijalva et al., "Letter Urging President Biden to Support a Precautionary Pause on Deep Seabed Mining," U.S. Congress, June 28, 2024, https://democrats-naturalresources.house.gov/imo/media/doc/letter_urging_president_biden_to_support_a_precautionary_pause_on_deep_seabed_mining.pdf.

485 Seaver Wang, "No, Collecting Seafloor Metals Won't Wreck the Ocean Carbon Cycle," Breakthrough Institute, July 9, 2024, https://thebreakthrough.org/issues/energy/no-collecting-seafloor-metals-wont-wreck-the-ocean-carbon-cycle.

486 Stu Robarts, "How Impossible Metals Plans to Mine the Seabed with Autonomous Robots," Mining Technology, May 29, 2024, https://www.mining-technology.com/interviews/how-impossible-metals-plans-to-mine-the-seabed-with-autonomous-robots/.

487 Caitlin Keating-Bitonti, *Seabed Mining in Areas Beyond National Jurisdiction: Issues for Congress*, CRS Report No. R47324 (Washington, DC: Congressional Research Service, 2022), 10, https://crsreports.congress.gov/product/pdf/R/R47324.

488 "The Mining Code," International Seabed Authority, accessed September 3, 2024, https://www.isa.org.jm/the-mining-code/.

489 Chris Pickens et al., "From What-If to What-Now: Status of

the Deep-Sea Mining Regulations and Underlying Drivers for Outstanding Issues," *Marine Policy* 169 (January 2024): 105967, https://doi.org/10.1016/j.marpol.2023.105967; and "About ISA," International Seabed Authority, accessed September 17, 2024, https://www.isa.org.jm/about-isa/.

490 "United Nations Convention on the Law of the Sea," opened for signature December 10, 1982, Article 161 § 8(c), https://www.un.org/Depts/los/convention_agreements/texts/unclos/unclos_e.pdf.

491 Kai Tabacek, "Five Nations Add to Growing Concerns about Deep Sea Mining at Key International Meeting," Greenpeace UK, August 1, 2024, https://www.greenpeace.org.uk/news/five-nations-add-to-growing-concerns-about-deep-sea-mining-at-key-international-meeting/.

492 Elizabeth Claire Alberts, "'Trust Needs to Be Rebuilt': Interview with Candidate to Head U.N. Seabed-Mining Authority," Mongabay Environmental News, July 4, 2024, https://news.mongabay.com/2024/07/trust-needs-to-be-rebuilt-interview-with-candidate-to-head-u-n-seabed-mining-authority/.

493 Baron Public Affairs, *Deep-Sea Minerals: The Next Arena of U.S.-China Competition* (Washington, DC: Baron Public Affairs, 2024), https://www.baronpa.com/library/deep-sea-minerals-the-next-arena-of-u-s-china-competition/.

494 International Seabed Authority, "Beijing Pioneer Hi-Tech Development Corporation Ltd"; and International Seabed Authority, "China Minmetals Corporation."

495 Michael Lodge, "The Regulation of Seabed Mining: A Conversation with Michael Lodge, Secretary General of the International Seabed Authority," moderated by Duncan Wood, Wilson Center, November 29, 2023, 33:30 to 33:35, https://www.wilsoncenter.org/event/regulation-seabed-mining-conversation-michael-lodge-secretary-general-international-seabed; and Asterios Tsioumanis and María Ovalle, "Summary of the Twenty-eighth Annual Session of the International Seabed Authority (Second Part): 10-28 July 2023," *Earth Negotiations Bulletin* 25, no. 253 (July 31, 2023), https://enb.iisd.org/sites/default/files/2023-07/enb25253e.pdf.

496 "Exploration Contracts," International Seabed Authority.

497 "Voices Calling for a Moratorium: Governments and Parliamentarians," Deep Sea Conservation Coalition, accessed September 18, 2024, https://deep-sea-conservation.org/solutions/no-deep-sea-mining/momentum-for-a-moratorium/governments-and-parliamentarians/.

498 John Negroponte et al., "Letter in Support of Senate Advice and Consent for the Convention on the Law of the Sea," PoliticoPro, March 11, 2024, https://subscriber.politicopro.com/f/?id=0000018e-3661-de4d-a19e-b76173430000.

499 Robert Beckman, "On the United States, the UN Convention on the Law of the Sea and US Freedom of Navigation Operations," *Fulcrum*, August 5, 2022, https://fulcrum.sg/on-the-united-states-the-un-convention-on-the-law-of-the-sea-and-us-freedom-of-navigation-operations/.

500 Ibid.; and The Law of the Sea Convention: Hearings Before the Senate Committee on Foreign Relations, 112th Cong., 2nd sess. (2012) (statement of Donald Rumsfeld, former U.S. Secretary of Defense), https://www.govinfo.gov/content/pkg/CHRG-112shrg77375/html/CHRG-112shrg77375.htm.

501 *Submerged Lands Act*, 43 U.S. Code (1953) §§ 1301 et seq., https://uscode.house.gov/view.xhtml?path=/prelim@title43/chapter29&edition=prelim; and Yusuf Khan, "Hawaii Bans Deep-Sea Mining as U.S. Political Support Splits on Party Lines," Wall Street Journal, July 9, 2024, https://www.wsj.com/articles/hawaii-bans-deep-sea-mining-as-u-s-political-support-splits-on-party-lines-2842ce05.

502 Internal Revenue Service, "Clean Vehicle Credits."

503 Benchmark Mineral Intelligence, *The Metals Company - Life Cycle Assessment for TMC's NORI-D Polymetallic Nodule Project and Comparison to Key Land-Based Routes for Producing Nickel, Cobalt and Copper* (London: Benchmark Mineral Intelligence, 2023), https://metals.co/wp-content/uploads/2023/03/TMC_NORI-D_LCA_Final_Report_March2023.pdf; and AMC Consultants Pty Ltd., *Technical Report Summary: TOML Mineral Resource, Clarion Clipperton Zone, Pacific Ocean, DeepGreen Metals Inc.* (Melbourne: AMC Consultants Pty Ltd., 2021), https://int.nyt.com/data/documenttools/2021-03-metals-company-technical-report-on-toml-mining-zone-plan/2d5350243bade994/full.pdf.

504 Niels Verbaan, "On Shore Processing of Polymetallic Nodules: A Review of Metallurgical Flowsheets Being Considered," presentation at Deep Sea Mining Summit, London, April 26, 2022, https://www.slideshare.net/iQHub/onshore-processing-of-nodules-a-review-of-metallurgical-flowsheets-and-active-testwork-updates.

505 "Qualifying Advanced Energy Project Credit (48C) Program," U.S. Department of Energy, accessed October 14, 2024, https://www.energy.gov/infrastructure/qualifying-advanced-energy-project-credit-48c-program; "Defense Production Act Title III," U.S. Department of Defense, accessed October 14, 2024, https://www.businessdefense.gov/ibr/mceip/dpai/dpat3/index.html; and "Loan Programs Office," U.S. Department of Energy, accessed October 14, 2024, https://www.energy.gov/lpo/loan-programs-office.

506 "America's Offshore Critical Mineral Resources," Bureau of Ocean Energy Management, November 15, 2019, https://www.boem.gov/sites/default/files/documents/marine-minerals/Critical%20Mineral%20State.pdf.

507 Mark G. Macklin et al., "Impacts of Metal Mining on River Systems: A Global Assessment," Science 381, no. 6664 (September 2023): 1345-50, https://www.researchgate.net/publication/374112762_Impacts_of_metal_mining_on_river_systems_a_global_assessment.

508 R. Anthony Hodge et al., "The Global Mining Industry: Corporate Profile, Complexity, and Change," Mineral Economics 35, no. 3-4 (December 2022): 587-606, https://link.springer.com/article/10.1007/s13563-022-00343-1.

509 Paul Manalo, "Discovery to Production Averages 15.7 Years for

127 Mines," S&P Global, June 6, 2023, https://www.spglobal.com/market-intelligence/en/news-insights/research/discovery-to-production-averages-15-7-years-for-127-mines.

510 Mohsen Bonakdarpour et al., "Mine Development Times: The US in Perspective," S&P Global, June 2024, https://cdn.ihsmarkit.com/www/pdf/0724/SPGlobal_NMA_Development-TimesUSinPerspective_June_2024.pdf.

511 Lawrence M. Cathles and Adam C. Simon, "Copper Mining and Vehicle Electrification," International Energy Forum, May 15, 2024, https://www.ief.org/focus/ief-reports/copper-mining-and-vehicle-electrification.

512 European Commission, *Regulation (EU) 2024/1252 of the European Parliament and of the Council of 11 April 2024 establishing a framework for ensuring a secure and sustainable supply of critical raw materials and amending Regulations (EU) No 168/2013, (EU) 2018/858, (EU) 2018/1724 and (EU) 2019/1020 (Text with EEA relevance)*, PE/78/2023/REV/1, May 3, 2024, https://eur-lex.europa.eu/legal-content/EN/TXT/PDF/?uri=CELEX:02024R1252-20240503.

513 Ibid.; and "Nature Restoration Law," European Commission, accessed October 30, 2024, https://environment.ec.europa.eu/topics/nature-and-biodiversity/nature-restoration-law_en.

514 Eleni Diamantopoulou, "Accelerating EU Mineral Permitting: The Critical Raw Materials Act," Womble Bond Dickinson, May 3, 2024, https://www.womblebonddickinson.com/uk/insights/articles-and-briefings/accelerating-eu-mineral-permitting-critical-raw-materials-act.

515 Eva Liedholm Johnson, Magnus Ericsson, and Anton Löf, "The Mining Permitting Process in Selected Developed Economies," *Land Use Policy* 131 (August 2023): 106762, https://www.sciencedirect.com/science/article/pii/S0264837723002284.

516 International Council on Mining and Metals, "Mining Principles," International Council on Mining and Metals, August 8, 2024, https://www.icmm.com/website/publications/pdfs/mining-principles/mining-principles.pdf?cb=83767.

517 "About us," ICMM, July 15, 2024, https://www.icmm.com/en-gb/our-story/who-we-are

518 "TSM 101: A Primer," Towards Sustainable Mining, July 11, 2024, https://mining.ca/wp-content/uploads/dlm_uploads/2024/07/TSM-PRIMER-2024-EN5.pdf; "The Copper Mark Theory of Change," The Copper Mark, July 15, 2024, https://coppermark.org/wp-content/uploads/2024/07/CopperMark_Impact_TheoryofChange_2024-07-15.pdf; and World Gold Council, *Responsible Gold Mining Principles* (London: World Gold Council, 2019), https://www.gold.org/download/file/14254/Responsible-Gold-Mining-Principles-en.pdf.

519 "Who We Are," ResponsibleSteel, accessed October 30, 2024, https://www.responsiblesteel.org/who-we-are; and "About ASI," Aluminum Stewardship Initiative, accessed October 30, 2024, https://aluminium-stewardship.org/about-asi/vision-mission-values.

520 "About the Initiative," Consolidated Mining Standard Initiative, accessed October 30, 2024, https://miningstandardinitiative.org/about-the-initiative/.

521 European Commission, "Critical Raw Materials: Ensuring Secure and Sustainable Supply Chains for EU's Green and Digital Future," press release, March 15, 2023, https://ec.europa.eu/commission/presscorner/detail/en/ip_23_1661.

522 "Responsible Sourcing," London Metal Exchange, July 15, 2024, https://www.lme.com/en/education/online-resources/lme-insight/lme-responsible-sourcing.

523 London Metal Exchange, "Discovering the Low Carbon Premium for the Nickel Market," press release, March 5, 2024, https://www.lme.com/api/sitecore/MemberNoticesSearchApi/Download?id=1dcae570-87ee-422a-9596-acaa2b041f11.

524 "What Are Scope 1, 2 and 3 Carbon Emissions?," National Grid, accessed October 30, 2024, https://www.nationalgrid.com/stories/energy-explained/what-are-scope-1-2-3-carbon-emissions.

525 BHP, "Western Australia Nickel to Temporarily Suspend Operations," press release, July 11, 2024, https://www.bhp.com/news/media-centre/releases/2024/07/western-australia-nickel-to-temporarily-suspend-operations.

526 "About the Community of Interest (COI) Panel," Mining Association of Canada, accessed October 30, 2024, https://mining.ca/towards-sustainable-mining/community-of-interest-advisory-panel/.

527 Steffany Bermúdez and Florencia Sarmiento, "Will the Inclusion of Voluntary Standards in Trade Agreements Lead to More Sustainable Trade?," International Institute for Sustainable Development, April 23, 2024, https://www.iisd.org/system/files/2024-04/ssi-voluntary-standards-agreements-sustainable-trade.pdf.

528 Janne Bemelmans et al., "Trade Effects of Voluntary Sustainability Standards in Tropical Commodity Sectors," *Food Policy* 118 (July 2023): 102440, https://www.sciencedirect.com/science/article/abs/pii/S0306919223000386.

529 Cristina Larrea et al., "Voluntary Sustainability Standards, Forest Conservation, and Environmental Provisions in International Trade Policy," International Institute for Sustainable Development, October 28, 2021, https://www.iisd.org/system/files/2021-10/voluntary-sustainability-standards-forest-conservation-trade-policy.pdf.

530 RMI, "Major Corporations Come Together to Advance the First Commercial Batch of Sustainable Steel in the US," press release, September 20, 2023, https://rmi.org/press-release/major-corporations-come-together-to-advance-the-first-commercial-batch-of-sustainable-steel-in-the-us/.

531 Keith Bradsher, "China Bans Rare Earth Exports to Japan amid Tension," CNBC, September 23, 2010, https://www.cnbc.com/2010/09/23/china-bans-rare-earth-exports-to-japan-amid-tension.html.

532 "Timeline: U.S.-China Relations," Council on Foreign Relations, accessed October 26, 2024, https://www.cfr.org/time-

line/us-china-relations.

533 Bradsher, "China Bans."

534 Ben Blanchard and Tom Miles, "China 2010 Rare Earth Exports Slip, Value Rockets," Reuters, January 19, 2011, https://www.reuters.com/article/business/environment/china-2010-rare-earth-exports-slip-value-rockets-idUSTRE70I11T/.

535 The White House, "Remarks by the President on Fair Trade," remarks, March 13, 2012, https://obamawhitehouse.archives.gov/the-press-office/2012/03/13/remarks-president-fair-trade.

536 CNN Wire Staff, "Obama Announces WTO Case against China over Rare Earths," CNN, March 13, 2012, https://www.cnn.com/2012/03/13/world/asia/china-rare-earths-case/index.html.

537 Office of the U.S. Trade Representative, "U.S. Trade Representative Michael Froman Announces U.S. Victory in Challenge to China's Rare Earth Export Restraints," press release, August 7, 2014, https://ustr.gov/about-us/policy-offices/press-office/press-releases/2014/August/USTR-Froman-US-Victory-in-Challenge-to-China-Rare-Earth-Export-Restraints.

538 Molycorp, Inc., "Molycorp, Inc. Signs Restructuring Support Agreement with Key Creditors," Mining.com, June 25, 2015, https://www.mining.com/web/molycorp-inc-signs-restructuring-support-agreement-with-key-creditors/.

539 Executive Office of the President, "Executive Order 13817 of December 20, 2017: A Federal Strategy to Ensure Secure and Reliable Supplies of Critical Minerals," *Federal Register* 82, no. 246 (December 20, 2017): 60835, https://www.federalregister.gov/documents/2017/12/26/2017-27899/a-federal-strategy-to-ensure-secure-and-reliable-supplies-of-critical-minerals.

540 U.S. Geological Survey, "Interior Releases 2018's Final List of 35 Minerals Deemed Critical to U.S. National Security and the Economy," press release, May 18, 2018, https://www.usgs.gov/news/national-news-release/interior-releases-2018s-final-list-35-minerals-deemed-critical-us.

541 U.S. Department of State, "Joint Statement on U.S.-Greenland MOU and Hyperspectral Survey," press release, June 6, 2019, https://2017-2021.state.gov/joint-statement-on-u-s-greenland-mou-and-hyperspectral-survey/index.html.

542 "Energy Resource Governance Initiative (ERGI)," U.S. Department of State, December 15, 2020, https://www.state.gov/wp-content/uploads/2020/12/Marketing-Materials_ERGI-One-Pager_2.20.20.pdf.

543 TechMet, "TechMet Receives Investment from US International Development Finance Corporation," press release, October 5, 2020, https://www.techmet.com/press-release-05-october-2020/; and U.S. Department of the Interior, "Trump Administration Announces Strategy to Strengthen America's Economy, Defense," press release, June 4, 2019, https://www.doi.gov/pressreleases/trump-administration-announces-strategy-strengthen-americas-economy-defense.

544 "Fact Sheet: Prioritizing Climate in Foreign Policy and National Security," The White House, October 21, 2021, https://www.whitehouse.gov/briefing-room/statements-releases/2021/10/21/fact-sheet-prioritizing-climate-in-foreign-policy-and-national-security/.

545 "Executive Order 14017 of February 24, 2021: America's Supply Chains," *Federal Register* 86, no. 38 (March 1, 2021): 11849, https://www.federalregister.gov/documents/2021/03/01/2021-04280/americas-supply-chains.

546 U.S. Geological Survey, "U.S. Geological Survey Releases 2022 List of Critical Minerals," press release, February 22, 2022, https://www.usgs.gov/news/national-news-release/us-geological-survey-releases-2022-list-critical-minerals.

547 "What Are Critical Materials and Critical Minerals?," U.S. Department of Energy, accessed October 26, 2024, https://www.energy.gov/cmm/what-are-critical-materials-and-critical-minerals.

548 International Energy Agency, *Global Critical Minerals Outlook 2024* (Paris: International Energy Agency, 2024), 108–123, https://www.iea.org/reports/global-critical-minerals-outlook-2024.

549 "Inflation Reduction Act and Bipartisan Infrastructure Law Resource Center," BlueGreen Alliance, accessed October 26, 2024, https://www.bluegreenalliance.org/site/inflation-reduction-act-and-bipartisan-infrastructure-law-resource-center/.

550 Clea Caulcutt, "Trade War Averted? Macron Gets Biden to 'Tweak' His Industrial Subsidies," *Politico* (EU edition), December 2, 2022, https://www.politico.eu/article/emmanuel-macron-joe-biden-us-france-lands-unexpected-concession-on-inflation-reduction-act/.

551 "Minerals Security Partnership," U.S. Department of State, accessed October 26, 2024, https://www.state.gov/minerals-security-partnership/.

552 "Office of the U.S. Special Coordinator for the Partnership for Global Infrastructure and Investment," U.S. Department of State, accessed October 26, 2024, https://www.state.gov/bureaus-offices/secretary-of-state/office-of-the-u-s-special-coordinator-for-the-partnership-for-global-infrastructure-and-investment/; Andres Schipani, "The US-Backed Railway Sparking a Battle for African Copper," *Financial Times*, August 21, 2024, https://www.ft.com/content/cb2823c7-f451-4bc9-959e-ec7e07384a31; and U.S. Agency for International Development, "The U.S. Government Announces Grant to Promote Investment Opportunities in the Democratic Republic of the Congo's Minerals Sector," press release, June 11, 2024, https://www.usaid.gov/democratic-republic-congo/press-release/jun-11-2024-us-government-announces-grant-promote-investment-opportunities-democratic-republic-congos-minerals-sector.

553 "TechMet Receives Further $50 Million Commitment from U.S. International Development Finance Corporation at COP28," Business Wire, December 1, 2023, https://www.businesswire.com/news/home/20231201346504/en/TechMet-Receives-further-50-Million-Commitment-From-U.S.-International-Development-Finance-Corporation-at-COP28.

554 Frank Fannon, "The United States Should Fix the FEOC Guidance," CSIS, *Commentary*, March 14, 2024, https://www.csis.org/analysis/united-states-should-fix-feoc-guidance.

555 "China's Curbs on Exports of Strategic Minerals," Reuters, August 15, 2024, https://www.reuters.com/markets/commodities/chinas-curbs-exports-strategic-minerals-2024-08-15/; and Reuters, "China to Limit Antimony Exports in Latest Critical Mineral Curbs," CNN, August 16, 2024, https://www.cnn.com/2024/08/15/tech/china-antimony-export-ban-intl-hnk/index.html.

556 Bloomberg News, "China Cobalt Buyers Use Global Glut to Challenge Pricing," Mining.com, January 16, 2024, https://www.mining.com/web/china-cobalt-buyers-use-global-glut-to-challenge-pricing/.

557 Arjun Murti, "Next 10 Years of Perpetual Transition: China," Super-Spiked, September 7, 2024, https://arjunmurti.substack.com/p/next-10-years-of-perpetual-transition.

558 Shannon Tiezzi, "With Kerry in Beijing, China Signals It Will Go Its Own Way on Climate Change Policy," *The Diplomat*, July 20, 2023, https://thediplomat.com/2023/07/with-kerry-in-beijing-china-signals-it-will-go-its-own-way-on-climate-change-policy/.

559 Daniel F. Runde, "Congress Should Support a 'Fix' to the Development Finance Corporation," *The Hill*, March 24, 2022, https://thehill.com/opinion/finance/599569-congress-should-support-a-fix-to-the-development-finance-corporation/.

560 Daniel F. Runde, "The U.S. EXIM Bank in an Age of Great Power Competition," Center for Strategic and International Studies, June 18, 2024, https://www.csis.org/analysis/us-exim-bank-age-great-power-competition.

561 "Permitting Process Flowchart for a Solar Project," The Permitting Institute, accessed October 26, 2024, https://www.permittinginstitute.org/solar-chart.

562 Wayne Winegarden, "Energy Transition or Not, Permitting Reform Is Essential," *Forbes*, June 19, 2024, https://www.forbes.com/sites/waynewinegarden/2024/06/17/energy-transition-or-not-permitting-reform-is-essential/; and U.S. Senate Committee on Energy and Natural Resources, "Manchin, Barrasso Release Bipartisan Energy Permitting Reform Legislation," press release, July 22, 2024, https://www.energy.senate.gov/2024/7/manchin-barrasso-release-bipartisan-energy-permitting-reform-legislation.

563 Ted Mann and Timothy Puko, "Senate Tax-and-Climate Plan Hinges on Streamlining Energy Projects," *Wall Street Journal*, July 31, 2022, https://www.wsj.com/articles/senate-tax-and-climate-plan-hinges-on-streamlining-energy-projects-11659276002?mod=e2tw; and "Let's Reform America's Clean Energy Permitting Process," Citizens' Climate Lobby, accessed October 26, 2024, https://citizensclimatelobby.org/our-climate-solutions/clean-energy-permitting-reform/.

564 U.S. Chamber of Commerce (@USChamber), "Why the U.S. Needs to Reform Its Permitting System," YouTube video, August 9, 2023, 30:54, https://www.uschamber.com/on-demand/infrastructure/common-grounds-us-needs-permitting-reform.

565 PRNewswire, "United States Ranks Next to Last in Development Time for New Mines That Produce Critical Minerals for Energy Transition, S&P Global Finds," S&P Global, July 18, 2024, https://press.spglobal.com/2024-07-18-United-States-Ranks-Next-to-Last-in-Development-Time-for-New-Mines-that-Produce-Critical-Minerals-for-Energy-Transition,-S-P-Global-Finds.

566 Philip Rossetti, "Manchin and Barrasso Revive Permitting Reform This Congress," R Street, July 30, 2024, https://www.rstreet.org/commentary/manchin-and-barrasso-revive-permitting-reform-this-congress/; and Herman K. Trabish, "Congressional Action on Energy Permitting Remains Stuck, but States, Developers Are Finding Solutions," Utility Dive, February 27, 2024, https://www.utilitydive.com/news/federal-energy-permitting-reform-doe-ferc-congress/705818/.

567 Earthjustice, "Ahead of Senate Markup, Earthjustice Blasts Manchin-Barrasso Permitting Bill," press release, July 31, 2024, https://earthjustice.org/press/2024/ahead-of-senate-markup-earthjustice-blasts-manchin-barrasso-permitting-bill.

568 "The China Threat," Federal Bureau of Investigation, accessed October 26, 2024, https://www.fbi.gov/investigate/counterintelligence/the-china-threat.

569 The White House, *Building Resilient Supply Chains, Revitalizing American Manufacturing, and Fostering Broad-Based Growth: 100-Day Reviews under Executive Order 14017* (Washington, D.C.: The White House, 2021). https://www.whitehouse.gov/wp-content/uploads/2021/06/100-day-supply-chain-review-report.pdf.

570 Chong-En Bai, Chang-Tai Hsieh, Zheng Michael Song, and Xin Wang, "The Rise of the State-Connected Private Sector in China," Stanford Center on China's Economy and Institutions, updated November 1, 2023, https://sccei.fsi.stanford.edu/china-briefs/rise-state-connected-private-sector-china.

571 Patsy Widakuswara, "Biden, EU Chief Downplay Differences over US Climate Subsidies," Voice of America, March 11, 2023, https://www.voanews.com/a/biden-to-host-eu-chief-to-discuss-china-climate-subsidies/7000145.html; and Peter Campbell and Harry Dempsey, "Chinese Battery Makers Set to Dominate Europe's Car Industry," *Financial Times*, December 6, 2022, https://www.ft.com/content/d407772c-4a76-4e59-9bb0-998b3f22383b.

572 Shannon Tiezzi, "China Celebrates German Chancellor's Trip," *The Diplomat*, November 5, 2022, https://thediplomat.com/2022/11/china-celebrates-german-chancellors-trip/.

573 Andrew Duehren, "U.S. and EU Advance Buyers' Club for EV Battery Minerals," *Wall Street Journal*, February 15, 2023, https://www.wsj.com/articles/u-s-and-eu-advance-buyers-club-for-ev-battery-minerals-5288287e.

574 John F. Kennedy, "Address to Joint Session of Congress May 25, 1961," (speech, Washington, D.C., May 25, 1961), https://www.jfklibrary.org/learn/about-jfk/historic-speeches/address-to-joint-session-of-congress-may-25-1961.

575 Strategic and Critical Materials Stock Piling Act, Public Law

76-117, US Statutes at Large 53 (1939): 811, as amended through Public Law 118-31 (2023).

576 "Strategic Materials Sales," Defense Logistics Agency, accessed January 6, 2025, https://www.dla.mil/Strategic-Materials/Sales/.

577 Gregory Wischer, "The U.S. Military and NATO Face Serious Risks of Mineral Shortages," Carnegie Endowment for International Peace, February 12, 2024, https://carnegieendowment.org/research/2024/02/the-us-military-and-nato-face-serious-risks-of-mineral-shortages?lang=en.

578 Hans Nicholas Jong, "Indonesian Nickel Project Harms Environment and Human Rights, Report Says," Mongabay Environmental News, February 26, 2024, https://news.mongabay.com/2024/02/indonesian-nickel-project-harms-environment-and-human-rights-report-says/; Amy Hawkins and Nina Lakhani, "China Accused of Scores of Abuses Linked to 'green Mineral' Mining," *The Guardian*, July 5, 2023, https://www.theguardian.com/world/2023/jul/05/china-mineral-mining-green-energy; and "Breaking: China-Congo 'Deal of the Century' Linked to Fraud, Bribery," The Sentry, November 29, 2021, https://thesentry.org/2021/11/28/6516/breaking-china-congo-deal-century-linked-fraud-bribery/.

579 "Workforce Trends in the U.S. Mining Industry," SME: Society for Mining, Metallurgy, and Exploration, accessed January 6, 2025, https://www.smenet.org/What-We-Do/Technical-Briefings/Workforce-Trends-in-the-US-Mining-Industry.

580 IEA, *The Role of Critical Minerals in Clean Energy Transitions: Executive Summary* (Paris: IEA, March 2022), https://www.iea.org/reports/the-role-of-critical-minerals-in-clean-energy-transitions/executive-summary.

581 "Could AI Help Find Valuable Mineral Deposits?," *The Economist*, November 1, 2023, https://www.economist.com/science-and-technology/2023/11/01/could-ai-help-find-valuable-mineral-deposits.

582 Josh Birenbaum and Elaine K. Dezenski, "Tightening the Belt or End of the Road? China's BRI at 10," FDD: Foundation for Defense of Democracies, February 27, 2024, https://www.fdd.org/analysis/2024/02/27/tightening-the-belt-or-end-of-the-road-chinas-bri-at-10/.

583 Seth G. Jones and Alexander Palmer, *China Outpacing U.S. Defense Industrial Base* (Washington, DC: CSIS, March 2024), https://www.csis.org/analysis/china-outpacing-us-defense-industrial-base.

www.ingramcontent.com/pod-product-compliance
Ingram Content Group UK Ltd.
Pitfield, Milton Keynes, MK11 3LW, UK
UKHW052012041026
15590UKWH00008B/2291

9 798765 198360